Insect Pests of Millets

Insect Pests of Millets
Systematics, Bionomics, and Management

A. Kalaisekar
P.G. Padmaja
V.R. Bhagwat
J.V. Patil
ICAR - Indian Institute of Millets Research
Hyderabad, India

AMSTERDAM • BOSTON • HEIDELBERG • LONDON
NEW YORK • OXFORD • PARIS • SAN DIEGO
SAN FRANCISCO • SINGAPORE • SYDNEY • TOKYO

Academic Press is an imprint of Elsevier

Academic Press is an imprint of Elsevier
125 London Wall, London EC2Y 5AS, United Kingdom
525 B Street, Suite 1800, San Diego, CA 92101-4495, United States
50 Hampshire Street, 5th Floor, Cambridge, MA 02139, United States
The Boulevard, Langford Lane, Kidlington, Oxford OX5 1GB, United Kingdom

Notices
Knowledge and best practice in this field are constantly changing. As new research and experience broaden our
understanding, changes in research methods, professional practices, or medical treatment may become necessary.

Practitioners and researchers must always rely on their own experience and knowledge in evaluating and using
any information, methods, compounds, or experiments described herein. In using such information or methods
they should be mindful of their own safety and the safety of others, including parties for whom they have a
professional responsibility.

To the fullest extent of the law, neither the Publisher nor the authors, contributors, or editors, assume any liability
for any injury and/or damage to persons or property as a matter of products liability, negligence or otherwise, or
from any use or operation of any methods, products, instructions, or ideas contained in the material herein.

Library of Congress Cataloging-in-Publication Data
A catalog record for this book is available from the Library of Congress

British Library Cataloguing-in-Publication Data
A catalogue record for this book is available from the British Library

ISBN: 978-0-12-804243-4

For information on all Academic Press publications
visit our website at https://www.elsevier.com/

Working together
to grow libraries in
developing countries

www.elsevier.com • www.bookaid.org

Publisher: Nikki Levy
Acquisition Editor: Nancy Maragioglio
Editorial Project Manager: Billie Jean Fernandez
Production Project Manager: Caroline Johnson
Designer: Victoria Pearson

Typeset by TNQ Books and Journals

We dedicate this book to all the millet-growing farmers of the world.

"Howe'er they roam, the world must follow still the plougher's team;
Though toilsome, culture of the ground as noblest toil esteem." (Thirukkural-1031)

–Thiruvalluvar (Tamil Philosopher, c. 2nd Century BC)

Contents

Preface

The Classic Maya Civilization in the lowlands of northern Guatemala collapsed mysteriously in the ninth century AD; sustained crop failure due to an epidemic of plant hopper (*Peregrinus maidis*) borne maize mosaic virus is proposed as a primary contributing cause of the collapse.

—Brewbaker (1979)

Insects are by far the greatest challenge to human food security and existence. There are several historical examples of great devastations ranging from mass migrations to even the collapse of human civilization settlements due to insect outbreaks. The plant bug (*P. maidis*)-orchestrated the demise of the Classic Maya Civilization in the AD 9th century and a series of transcontinental locust outbreaks in the recent past are some of the archetypical examples of the destructive ability of insects. In Africa and central Asia, during 2003–05, an estimated US$500 million was spent on taming the onslaught of a locust plague. Even political instabilities seem to abet the pest outbreaks in some instances. The 2012 locust swarm in north African countries was articulated as "the fall of Gaddafi leading to the rise of desert locusts" and this was acknowledged by the FAO.

Plants and insects radiated out to several lineages as a means of coevolution throughout the geological timescale. Artificial cultivation of plants for food at the dawn of human civilization brought in a new dimension to the millions of years old natural plant–insect association. Cultivated plants became the most preferred food courts for several new radiants of pestiferous insect guilds owing to the lack of natural selection against the herbivorous insect community. Insect pests swiftly respond to even a slight resistance in cultivars of crop plants by producing local populations or biotypes. Therefore, virtually any attempt against insects, either in the form of human interference or by way of plants themselves developing resistance, is decisively neutralized by insect pests. Such resilient capacity of insects leads to increased pest pressure especially on crop plants.

Millets as a group of cultivated plants are no stranger to the insect pests. There are at least 450 insect species recorded globally on millets. The production economy of millets does not allow the farmer to take up cost-intensive pest control measures as the crops are cultivated in resource-deprived farms. Thus, the cost–benefit conundrum of pest control on one side and the ability of insects to overcome pest control attempts on the other side forms a critical combination in the millet production system. Such a special condition warrants a thorough understanding of the entire gamut of key insect pests associated with the millet ecosystem. We, in this volume, brought together all the scattered relevant research findings to provide a comprehensive understanding of insect pests of millets. This book is uniquely designed to provide all the available information on insects associated with millets in a lucid manner. The book also contains color images of insect pests, damage symptoms, and diagrammatic explanations to help the reader in identifying the pests. We intend the book to benefit a wide realm of readership engaged in research, teaching, farming, extension, and pest control, in addition to students.

Authors

Acknowledgments

We thankfully acknowledge the numerous researchers whose contributions helped in the making of this book. We are ever grateful to Dr. Vilas A. Tonapi, Director, ICAR— Indian Institute of Millets Research, Hyderabad, India, for his support and guidance. We are thankful to our colleagues from IIMR, Drs. G. Shyam Prasad, B. Subbarayudu, and K. Srinivasa babu. We thank Ms. D. Roopa for her help in collecting and checking references. We thank Ms. Nancy Maragioglio, Ms. Billie Jean Fernandez, Ms. Caroline Johnson, and Ms. Victoria Pearson, all from Elsevier for their consistent support and patient pursuit throughout the process of manuscript preparation and finalization. The first author would like to express heartfelt gratitude to his parents Mr. M. Andiappan and Mrs. A. Kalaiselvi, and to his wife Mrs. K. Kanimozhi, daughter Durga and son Vishnu Priyan, for their patience and moral support extended throughout the period of preparation of this book.

Authors

INTRODUCTION

The first part of this chapter deals with basic aspects of millets; and the second part covers insect pests by crop and their importance as pests in millets along with a brief review of the nature of the damage they cause.

MILLET CULTIVATION: HISTORY, SCOPE, STATUS, AND PROSPECTS

A group of 35 domesticated grass species belonging to the botanical family Poaceae includes the world's staple food-producing crop plants that are functionally classified as cereals and millets. Millets are annual small-grained cereals grown in warm-climate regions of the world. Normally, the word "millet" refers to pearl millet and small millets such as finger millet, proso millet, kodo millet, foxtail millet, barnyard millet, little millet, etc. Nevertheless, sorghum also possesses the qualities of millets and for all practical purposes sorghum is considered a coarse-grained millet and is also known as great millet. In this book, reference to millets always includes sorghum.

Cultivation of millets is as old as the beginning of sedentism and civilization in the anthropological history of the world that dates back to around 8000 BC. It is believed that sorghum, finger millet, and pearl millet were of African origin, whereas foxtail millet, proso millet, and kodo millet were Asian in origin. Archeological evidence suggests that foxtail millet and proso millet are the oldest of the cultivated millets, even older than rice. It is interesting to note a BBC news report that the oldest noodles unearthed in China were made from millets 4000 years ago! Evidently the millets of Asian origin were under cultivation in India during 2500–1500 BC, whereas the African millets reached India around the same period. Pliny stated that sorghum entered Rome from India and spread to Europe. Fossil evidence proves that millets were in use as food even during prehistoric periods in India, China, and Africa. Recorded history starts from the ancient period and the earliest written evidence documents that millets were grown in most parts of the known world by then. During the medieval period, millets had become the principal food of the poor especially in Europe. Probably, this was the beginning of when millets were looked upon as poor man's cereal. The advent of yeast-raised bread made out of wheat and the higher yields achieved in crops like wheat, rice, corn, rye, and potato brought an eclipse of millet production in Europe during the 19th century. The other notable reason for the comparatively lesser preference of millets was their strong taste, the most preferred cereals, such as wheat and rice, having a bland and mild taste. The nutritional superiority of millets over cereals was downplayed with preference for milder taste. However, in other parts of the world, especially in India, China, and Africa, millets continued to be the main food source at least until the first half of the 20th century. Thereafter, even in Asia and Africa, millets witnessed a significantly smaller rate of increase in area and production compared to wheat, rice, and other cash crops. Area and

Insect Pests of Millets. http://dx.doi.org/10.1016/B978-0-12-804243-4.00001-X

production showed a declining trend in many parts of the world, especially after the late 1980s, mainly due to competitive crops replacing millets even in traditional areas.

Millet cultivation is now largely restricted to the semiarid tropics of the world, owing to the hardiness of the crops. At present the total cultivated area under millets (31.3 million ha), including sorghum (44.2 million ha), is over 75 million ha and the total production (millets: 27.8 million ton; sorghum: 67.8 million ton) is over 95 million tons as per FAO (2014) records. The leading millet producers of the world are India, Nigeria, Niger, and China. As far as sorghum is concerned, half of the total global grain produced is utilized for human food and the remaining half used for animal feed. Around 95% of other millets are cultivated mainly in India, China, Nigeria, and Niger, and the entire quantity is consumed as food locally, with a small proportion used as bird and animal feed. Almost the entire quantity of sorghum and other millets in the Americas, Europe, and Australia is put into animal feed and nonfood uses.

The declining area under millets and not so impressive production trend over the recent past certainly need to be addressed globally. The millet production system warrants global attention for the following reasons: first, global warming could lead to water scarcity and an increase in the frequency of drought, increased risk of heat or drought stress to crops and livestock, a likely change in the length of the growing period by way of accelerated physiological development that could hasten maturation and reduce yields, and increased respiration resulting in reduced yield potential. Millets, with their inherent ability to withstand such climatic vagaries, would be the harbinger of global food security. Millet plants are highly adaptable to climatic anomalies and can produce more biomass than other crops. These grasses are armored with C_4 photorespiratory physiology, which enables the plants to more efficiently utilize CO_2 and thrive under moisture stress. Second, the nutritional superiority of millets over cereals warrants immediate attention in the wake of alarmingly increasing lifestyle-induced health problems. The millet-based diet would surely provide substantial relief from such maladies. Fat in millets constitutes higher proportions of PUFAs. A higher content of unavailable carbohydrates (dietary fiber) and higher satiety effect (feeling of fullness) make the millets an ideal diet for healthy living. The slow release of sugars from a millet-based diet could make it a special food for diabetics and the obese. Sorghum bran possesses more antioxidants and antiinflammatory properties than blueberries and pomegranates. Millets are good sources of many essential minerals. Millets are gluten free and therefore could be an effective alternative for wheat and other such cereals for people with celiac complaints. There are many other considerations for promoting millets, such as uses in food (including health foods), feed, fodder (dry and green), and industrial raw materials (including bioethanol).

There are two sides to the causes of decline, namely the supply side and the demand side. The supply-side factors that pull the area down are marginalized cultivation, more remunerative crop alternatives, low profitability–low prices, lack of incentives, and decline in production and quality. Demand-side factors influencing the reduction are changing consumer tastes and preferences, rising per-capita income and social status, government policies, rapid urbanization, low shelf life, storage and inconvenience in food preparation, industrial uses not on a scale to encourage production, and minimal documentation of nutritional and health benefits.

To face all these challenges poised against the millet production system, a three-pronged strategy needs to be put in place, namely, focused research and development, government support, and awareness creation. Research and development efforts need to focus on varieties and hybrids with better regenerative capacity after drought, with tolerance to salinity and alkalinity, and that meet the requirements of market and industry; biofortification with micronutrients; addressing the problem of seed setting in pearl millet; validation of technologies in real farming situations; development of machinery to reduce drudgery, e.g., dehusking of small millets; improvement of shelf life; and scientific data on water use efficiency, agroclimatic limitations, and high nutritive values. Further, there is a need for sensitizing local governmental institutions to better percolation of millet promotion schemes. Awareness creation among various stakeholders requires emphasis on the following areas: exploring health benefits and nutritional advantages of millets and spreading the knowledge, entrepreneurship in value addition, exploring options for export of value-added millet products, and use of millets in the livestock and poultry feed industry. Further, consumer food products are fast moving out of traditional bases like wheat. The increasing presence of millet-based consumer food products like biscuits on the shelves of multibrand retailers is an encouraging market trend in favor of millets. Fast-growing health consciousness coupled with ever-increasing consumerism, especially in developing nations, indicates a future increase in demand for millet-based food products. Therefore, intensive millet cultivation is in the offing, and this could bring in several production constraints, such as increased pest problems.

INSECT PESTS

Insect pests are one of the major impediments in millet production systems across many areas in Africa and Asia. There has been a misconception that insect pests are not the major yield-reducing factor in millets. This is mainly due to two reasons: first, there have been no scientific yield-loss studies with reference to insect pest damage in millets. Second, millet production is being practiced as a subsistence and marginal enterprise in which loss due to insects gets little or no attention. Insect pests have certainly become major impediments wherever intensive and extensive cultivation is practiced. For example, sorghum and pearl millet are cultivated across several areas in India and Africa, respectively, and these areas are facing severe insect pest problems. The literature on specific details of insect pests feeding on millets is scanty. Except for some reviews on insects associated with millets (Sharma and Davies, 1988), small millet (Murthi and Harinarayana, 1989), and pearl millet (Harris and Nwanze, 1992), the majority of the literature comprises preliminary observational reports. There are some reports on localized insect outbreaks, especially in pearl millet-growing areas of India and Africa (Sharma and Davies, 1988).

The available literature abounds with synonyms being counted and reported as separate entities. Thus, the number of pests reported on millets goes far beyond the actual one. There is a separate chapter in this book on systematics and taxonomy to completely eliminate the multiplicity of names for the same pest due to the many synonymies being wrongly recognized as separate taxa and to give authentic, updated, valid nomenclature for all the insect pests reported on millets. In this book, all the major pests are dealt with in detail and the minor pests are also covered with relevant information.

SORGHUM, *SORGHUM BICOLOR*

(Syn. *Sorghum vulgare*)

Sorghum originated in Africa and is usually cultivated in tropical, subtropical, and arid regions of the world. It is a staple cereal food in many African and Asian countries and the interest in sorghum as a food source is increasing in many countries (Waniska and Rooney, 2000). Sorghum ranks fifth among the world's most important crops. Its current world production stands at 67.8 million tons (FAO, 2014).

Pest problems in sorghum start right at the seed stage and continue until harvest. There are around 150 insect pests recorded on sorghum globally. In India, the major insect pests of sorghum are shoot fly, stem borer, shoot bug, aphids, and a complex of earhead pests such as head bug and grain midge (Jotwani and Young, 1971). Other notable pests are white grubs, cutworms, grasshoppers, and leaf-eating beetles.

Root Feeders

White grubs, *Holotrichia serrata* and *Lachnosterna consanguinea*, sporadically assume serious pest status in Africa and India. The grubs feed on the roots of seedlings as well as older plants, resulting in withering of the plants. The infested plants wither and wilt in patches. Wireworms (Elateridae, Tenebrionidae: Coleoptera) also cause similar damage in sorghum seedlings. But the damage is generally not seen to affect the plants in patches, unlike white grubs. There are underground burrowing bugs, *Stibaropus* species, found sucking the sap from roots.

Termites, *Odontotermes* spp. and *Microtermes* sp., feed on the roots in some areas with sandy loam soils. Under dry conditions, termites feed on aerial parts of the plant also (Fig. 1.1). Damage to germinating seeds by ants, especially *Monomorium salomonis* and *Pheidole sulcaticeps*, affects plant populations.

Seedling Pests

The shoot fly, *Atherigona soccata,* is strictly a seedling pest. It causes damage to the seedlings of 1 week to 30 days of age. The typical symptom of damage is drying of the central shoot, called "deadheart" (Fig. 1.2). It is a major pest of economic importance in sorghum-growing areas of Asia, Africa, and Europe (Nwanze et al., 1992).

The cutworm, *Agrotis ipsilon*, damages plants by cutting the seedling at or a little below the soil surface, resulting in plant withering and lodging. Many a time the damage by cutworms gives an appearance of ruminant-animal grazing.

FIGURE 1.1

Termite damage in sorghum.

Courtesy of A. Kalaisekar.

FIGURE 1.2

Deadheart damage by *Atherigona soccata* in sorghum.

Courtesy of A. Kalaisekar.

Stem Borers and Leaf Feeders

There are seven stem borers (see Chapter 2) that cause economic loss in sorghum and among them the spotted stalk borer, *Chilo partellus*, and the pink borer, *Sesamia inferens*, are the most important in India and in Africa. The stem borer attack usually starts on 1-month-old crops and lasts until harvest.

The initial stage of crop growth, that is, from a little less than a month to up to 2 months, produces deadheart symptoms due to the internal feeding of the larvae. The initial instar larvae feed on leaves by making leaf scrapings and irregular holes (Fig. 1.3). Many times, the larval entry into the plant stem makes a horizontal series of "window holes" as the inner leaf whorl unfurls during the plant growth. The main symptomatic difference between *C. partellus* and *S. inferens* is that the stem tunneling damage by the former is seen with dried-up waste materials found within the leaf-sheath enclosure (Fig. 1.4). More extensive stem tunneling and the presence of yellowish jelly materials inside the leaf-sheath coverings can be seen with the latter (Fig. 1.5). The armyworm (Fig. 1.6) *Mythimna separata* causes damage by defoliation (Kundu and Kishore, 1971; Sharma et al., 1982; Sharma and Davies, 1988) and severely damaged plants are left with only the leaf midrib (Fig. 1.7). The larvae are gregarious and feed mostly at night. *Spodoptera exempta* causes severe defoliation in Africa and the adult moths possess the ability to fly hundreds of kilometers (Winch, 2006).

Sometimes the red hairy caterpillar *Amsacta moorei* causes severe defoliation (Srivastava and Goel, 1962). Occasional leaf-feeding insects are the caterpillars *Amsacta albistriga*, *Amsacta lactinea*, *Euproctis virguncula*, *Cnaphalocrocis patnalis*, and *Mocis frugalis*; the chrysomelid leaf beetles such as *Chaetocnema indica*, *Longitarsus* spp., and *Phyllotreta chotonica* usually make "shot holes" in the leaves. The ash weevil *Myllocerus undecimpustulatus maculosus* causes damage to sorghum by feeding on the foliage and, under severe infestations, the entire leaf blade is eaten up, leaving only the midribs (Kishore and Srivastava, 1976).

A host of grasshoppers, *Nomadacris septemfasciata*, *Acrida exaltata*, *Aiolopus longicornis*, *Aiolopus simulatrix*, *Aiolopus thalassinus*, *Atractomorpha crenulata*, *Chrotogonus hemipterus*,

FIGURE 1.3

Leaf-feeding damage by *Chilo partellus* in sorghum.

Courtesy of A. Kalaisekar.

FIGURE 1.4

Stem tunneling by *Chilo partellus* in sorghum.

Courtesy of A. Kalaisekar.

FIGURE 1.5

Stem tunneling by *Sesamia inferens* in sorghum.

Courtesy of A. Kalaisekar.

FIGURE 1.6

Larvae of *Mythimna separata*.

Courtesy of A. Kalaisekar.

Diabolocatantops axillaris, *Hieroglyphus banian*, and *Hieroglyphus nigrorepleptus*, attack sorghum at all stages of the crop. The feeding by both nymphs and adults causes moderate to severe defoliation. There are two grasshopper species, namely *Conocephalus maculatus* (Fig. 1.8) and *H. nigrorepleptus* (Fig. 1.9), that have become highly serious defoliators in eastern India and western India, respectively.

FIGURE 1.7

Defoliation by armyworm in sorghum.

Courtesy of A. Kalaisekar.

FIGURE 1.8

Defoliation by *Conocephalus maculatus* in sorghum.

Courtesy of A. Kalaisekar.

Sucking Pests

The shoot bug *Peregrinus maidis* is the major sucking pest in sorghum. Nymphs and adults feed in the leaf whorls during the rainy and postrainy seasons. Feeding causes leaf chlorosis and reddening (Fig. 1.10), and in severe cases the entire plant dries up. The shoot bug transmits two viral diseases, viz., maize mosaic virus (MMV) and maize stripe virus (MStpV), in sorghum and maize. *P. maidis* is the sole vector of MMV and MStpV, and the *Sorghum* spp. are the ancestral plant hosts (Nault, 1983).

FIGURE 1.9

Defoliation by *Hieroglyphus nigrorepleptus* in sorghum.

Courtesy of A. Kalaisekar.

FIGURE 1.10

Leaf chlorosis symptom by shoot bug in sorghum.

Courtesy of A. Kalaisekar.

FIGURE 1.11

Damage by *Melanaphis sacchari*.

Courtesy of A. Kalaisekar.

There are two important aphids that infest sorghum. The corn aphid *Rhopalosiphum maidis* is an important pest in sorghum and colonizes in large numbers inside leaf whorls, on the upper side of leaves, on the leaf sheath, and on earheads. *R. maidis* is a vector for maize dwarf mosaic virus. The sugarcane aphid *Melanaphis sacchari* usually sucks sap from the underside of the lower leaves (Fig. 1.11).

The sugarcane leafhopper *Pyrilla perpusilla* is a minor pest that sometimes assumes damaging proportions on sorghum. Several species of hemipterous bugs, *Cletus punctiger*, *Dolycoris indicus*, *Empoasca flavescens*, *Lygaeus* spp., *Menida histrio*, *Nephotettix virescens*, and *Nezara viridula*, are commonly occurring insect pests on sorghum. Though the bugs are almost always present on sorghum crops, they seldom make any economic loss. However, *N. viridula* and *D. indicus* sometimes could cause some loss.

Some thrips, especially *Caliothrips indicus*, *Sorghothrips jonnaphilus* are reported on sorghum (Ananthakrishnan, 1973).

Earhead Pests

The sorghum midge *Stenodiplosis sorghicola* is a cosmopolitan pest, found throughout the sorghum-growing areas of the world (Harris, 1976; Sharma et al., 1988). Maggots feed on the ovaries, leading to formation of chaffy florets (Fig. 1.12). The puparial cases adhered to the grains in the affected panicle can be seen as a confirmation of midge damage. Chaffy grains showing minute but visible circular holes are indications of the emergence holes of parasitoids of the midge.

The earhead bug *Calocoris angustatus* is an important earhead pest of sorghum causing considerable damage. Nymphs and adults colonize in large numbers on the earheads and suck the milky sap from developing grains. The whole earhead becomes dusty black and chaffy as a result of head bug damage (Fig. 1.13). Sometimes other bugs, like *Dysdercus koenigii* and *N. viridula*, also damage the developing grains.

Many species of lepidopteran caterpillars are found feeding on developing grains: *Autoba silicula*, *Cryptoblabes gnidiella*, *Cydia* spp., *Conogethes punctiferalis*, *Ephestia cautella*, *Eublemma* spp.,

FIGURE 1.12

Damage by grain midge in sorghum panicle.

Courtesy of A. Kalaisekar.

FIGURE 1.13

Damage by head bug in sorghum panicle.

Courtesy of A. Kalaisekar.

Euproctis limbata, Euproctis subnotata, Helicoverpa armigera. The maize cob borer *Stenachroia elongella* causes severe damage to earheads of sorghum especially in east India. The caterpillars cause extensive webbing of grains in the earhead and feed on the grains.

Some beetle species, *Chiloloba acuta, Mylabris pustulata*, and *Cylindrothorax tenuicollis*, are also found on earheads, especially during anthesis.

PEARL MILLET, *PENNISETUM TYPHOIDES*

(Syn. *Pennisetum americanum*, *Pennisetum glaucum*, *Pennisetum spicatum*)

Pearl millet, or bulrush millet, is the most widely grown among the millets in the semiarid tropics. Pearl millet is grown mainly for grains in west Africa, parts of east Africa, and the Indian subcontinent. In South Africa, the United States, and Australia, the crop is grown for fodder. Pearl millet is a major staple food in the semiarid regions of Africa and Asia. People in northern Namibia, many countries in the Sahel of Africa, and some areas in Rajasthan (India) are almost entirely dependent on pearl millet for food.

Insect pests cause considerable economic damage to pearl millet in India and Africa. Pearl millet can withstand acidic soil conditions. Pearl millet is more of a drought-avoiding crop than a drought-resistant one, as the plant grows faster and requires less water (Winch, 2006). Globally there are about 140 species of insects recorded on pearl millet (Sharma and Davies, 1988; Harris and Nwanze, 1992).

Seedling Pests

The shoot fly *Atherigona approximata* is the major seedling pest. Other shoot flies occasionally found on pearl millet are *A. soccata*, *Atherigona oryzae*, *Atherigona punctata*, *Atherigona ponti*, and *Atherigona yorki*.

The anthomyiid flies *Delia arambourgi* and *Delia flavibasis* are found to cause deadheart in Ethiopia. Larvae of some species of chloropid flies are found in the deadhearts (Deeming, 1971), but their role as damage-causing pests is questionable. The presence of chloropid larvae could be a secondary manifestation on the already rotten plant tissues created by the feeding of *Atherigona* species.

Stem Borers

There are as many as 22 lepidopteran borers recorded on pearl millet in Africa and India (Sharma and Davies, 1988) and the major ones are *Chilo partellus*, *C. infuscatellus*, *Sesamia calamistis*, *Sesamia cretica*, *S. inferens*, *Diatraea grandiosella*, *Coniesta ignefusalis*, *Busseola fusca*, *Eldana saccharina*, *Ostrinia furnacalis*, and *Ostrinia nubilalis*. *C. partellus* is a serious problem in India, whereas *C. ignefusalis*, *B. fusca*, and *S. calamistis* are the most important yield-reducing borers in Africa. The borer damage usually starts at 1-month crop and continues up to the grain maturation stage. The damage symptoms are similar to those found in sorghum.

Leaf Feeders

Lepidopteran caterpillars, viz., *Amsacta moorei*, *Mythimna loreyi*, *M. separata*, *Cnaphalocrocis medinalis*, *C. patnalis*, *Spodoptera exigua*, *Spodoptera frugiperda*, *Spodoptera mauritia*, *Autoba silicula*, and grasshoppers such as *Diabolocatantops axillaris*, *Hieroglyphus banian*, *H. daganensis*, *Oedaleus senegalensis*, and *Schistocerca gregaria* feed on foliage and sometimes cause severe damage. The skipper butterfly *Pelopidas mathias* is an occasional pest and longitudinally folds the leaf to feed from within (Fig. 1.14).

Sucking Pests

Bugs like *Blissus leucopterus*, *Nysius niger*, and hoppers, viz., *Cicadulina mbila*, *C. storeyi*, and *Pyrilla perpusilla* cause damage to the overall vigor of the plant stand. Aphids like *R. maidis*, *Sitobion miscanthi*, and *Hyalopterus pruni* occur on leaves and earhead causing discoloration, curling, and poor seed set.

Other Pests

The beetle *Chiloloba acuta* is found in large numbers on panicles, especially during anthesis. The millet midge *Geiromiya penniseti* is an important earhead pest causing considerable damage to pearl millet. *Euproctis* spp. (Fig. 1.15), *Orvasca subnotata*, and *Helicoverpa armigera* feed on panicles. *S. exempta*, *Spilarctia obliqua*, *Pachnoda interrupta*, *Phyllophaga* spp., *Thrips hawaiiensis*, *Myllocerus* spp., and *Peregrinus maidis* feed on vegetative parts of the plant. *Holotrichia consanguinea* and *H. serrata* are also reported as pests on roots of pearl millet.

FIGURE 1.14

Larva of *Pelopidas mathias*.

Courtesy of A. Kalaisekar.

FIGURE 1.15

Damage by larvae of *Euproctis similis* on pearl millet.

Courtesy of A. Kalaisekar.

FINGER MILLET, *ELEUSINE CORACANA*

Finger millet is also called African millet. It is widely cultivated and an important staple food grain in Africa. The other important producer of finger millet is India. Finger millet seeds can remain viable up to 10 years and therefore the grains can be "famine reserve seeds." The grains are less susceptible to birds in the field and to insects in storage. The finger millet crop is attacked by many insect pests especially under irrigated conditions. The cultivation is normally as a transplanted crop under irrigated conditions. There are at least 120 insect pest species recorded on finger millet in Asia and Africa.

Root Feeders

The root aphid *Tetraneura nigriabdominalis* is a serious pest in finger millet. The affected plants show withering symptoms initially and finally dry up. The white grub *Phyllophaga* sp. also sometimes causes considerable crop loss.

Shoot and Stem Feeders

Shoot fly infestations are generally rare in this crop. There are some reports of occurrences of *Atherigona miliaceae* and *A. soccata*. The damage symptoms are seen as deadheart. Shoot fly infestation usually occurs in seedlings of less than 1 month of age.

Among the stem borers *S. inferens* is the most serious pest in India and in Africa. *S. inferens* readily accepts finger millet as its preferred host plant rather than any other cereal or millet. The borer and the host plant are said to have originated in Africa and this could possibly be the reason for the preference. The infestation usually occurs in crops of more than 1 month. The affected plants produce deadheart symptoms (Fig. 1.16), usually with bore holes in the stem at the vegetative stage, and produce white-ear at the panicle stage (Fig. 1.17).The other borers, such as *C. partellus*, *B. fusca*, and *Saluria inficita*, also occur in finger millet.

The stem weevil *Listronotus bonariensis* is a minor pest in this crop. Adults feed on leaves and larvae feed inside the stems and cause drying of tillers. Under severe infestations, whiteheads, stem break, and lodging are the symptoms. In some areas in India, the infestation occurs frequently and is often mistaken for shoot fly or for lepidopteran borer attack. In wild relatives of finger millet such as *Eleusine indica*, the weevil infestation is very common.

Leaf Feeders

Hairy caterpillars, *Amsacta albistriga*, *A. transiens*, and *A. moorei*, feed on leaves and cause defoliation. Cutworms, *Agrotis ipsilon*, graze on the small seedlings at ground level. Armyworm larvae of *Spodoptera exempta*, *S. mauritia*, and *Mythimna separata* feed on leaves during the night and in severe infestations cause complete defoliation leaving the midrib. Leaf-folder *C. medinalis* larvae fold and feed on leaves. Skipper *P. mathias* larvae defoliate.

Grasshoppers, *Chrotogonus hemipterus*, *Nomadacris septemfasciata*, and *Locusta migratoria* nymphs and adults defoliate. Grubs of the beetle *Chnootriba similis* skeletonize the leaf. Thrips, *Heliothrips indicus* and other thrips species, feed by making whitish patches on leaves, especially at the seedling stage. Thrip damage is a common sight on the seedlings, almost invariably at the second to the fifth leaf stage in peninsular India.

FIGURE 1.16

Deadheart damage symptom by *Sesamia inferens* in finger millet.

Courtesy of A. Kalaisekar.

FIGURE 1.17

White-ear damage symptom by *Sesamia inferens* in finger millet.

Courtesy of A. Kalaisekar.

Sucking Pests

Aphids, *Hysteroneura setariae*, *Metopolophium dirhodum*, *R. maidis*, and *S. miscanthi*, are important pests and damage the plant from the seedling to the panicle stage. Mealy bug, *Brevennia rehi*, damage causes drying of the whole plant in patches. Leaf hoppers *Cicadulina bipunctella bipunctella* and *Cicadulina chinai* suck the plant sap, and severe infestation causes the whole plant to wither and dry up.

FOXTAIL MILLET, *SETARIA ITALICA*

(Syn. *Panicum italicum*)

The panicle looks like the tail of a fox and hence the crop is called foxtail millet. Foxtail millet is commonly known as Italian millet. It is one of the oldest domesticated crops. Foxtail millet is also a drought-avoiding crop like pearl millet by virtue of its fast-growing nature. It is mainly grown in India, Japan, China, southeast Europe, North Africa, and America (Winch, 2006). It grows very fast and matures in 75–80 days. This could be one of the reasons it escapes many common cereal insect pests. There are around 70 insect pests recorded on foxtail millet.

The major pest is the shoot fly *Atherigona atripalpis*, which produces deadheart symptoms (Fig. 1.18). Shoot flies are a major cause of concern especially in many small millets and need a special research focus (Jotwani et al., 1969; Jagadish, 1997; Jagadish and Neelu, 2006). Occasionally other species such as *Atherigona approximata*, *A. pulla*, *A. punctata*, and *A. biseta* have also been found infesting the crop. Shoot fly damage is seen even in the grown-up plants of more than 1 month, unlike in sorghum. It is a major pest of economic importance in sorghum-growing areas of Asia, Africa, and Europe.

The cutworm *Agrotis ipsilon* occasionally causes severe damage under dry conditions during the seedling stage of the crop. The gregarious larvae cut the whole seedlings in patches so as to produce an appearance of grazing by ruminants (Fig. 1.19).

The stem borer *C. partellus* causes considerable damage in some areas of India. Pink borer, *S. inferens*, and corn borer, *Ostrinia furnacalis*, also occasionally attack the crop. Stem borer attack usually starts on the 1-month-old crop and continues until harvest (Kundu and Kishore, 1971).

Armyworms, *M. separata*, *Spodoptera frugiperda*, and *S. litura*, occasionally cause severe damage by completely defoliating the leaves.

Occasional leaf-feeding caterpillars on foliage are *Amsacta albistriga*, *A. moorei* and *A. lactinea*; the ash weevil *M. undecimpustulatus maculosus* also cause damage.

The leaf beetle *Oulema melanopus* and flea beetle *Chaetocnema basalis* make shot holes on leaves.

The leaf folder *C. medinalis* and leaf roller *C. patnalis* sometimes occur and feed on leaves by making leaf rolls. Surface grasshoppers *Chrotogonus hemipterus* generally cause minor leaf damage. The grasshopper *Conocephalus maculatus* feeds on leaves as well as panicles. The green bug *N. viridula* is a serious pest on earheads and sucks the sap from developing grains resulting in chaffy grains. Several species of bugs, *Cletus punctiger*, *Dolycoris indicus*, and *Nephotettix virescens*, are reported. Sometimes the aphid *M. sacchari* occurs on this crop. The sugarcane leafhopper *Pyrilla perpusilla* is a minor pest.

FIGURE 1.18

Deadheart symptom by shoot fly in foxtail millet.

Courtesy of A. Kalaisekar.

FIGURE 1.19

Grazing symptom by cutworm in foxtail millet.

Courtesy of A. Kalaisekar.

KODO MILLET, *PASPALUM SCROBICULATUM*

P. scrobiculatum is found in wild (var. *commersonii*) form mostly in Africa and in cultivated (var. *scrobiculatum*) form in India. The crop is said to be still under the process of domestication as the wild and the cultivated forms cross-pollinate. The wild forms are generally found along paths and ditches and on disturbed ground. Therefore it is also called "ditch millet." The plant has a strong fibrous root system and can be a good cover crop in areas that are prone to soil erosion. Kodo millet had been the main staple food in south India, especially in Tamilnadu, until the late 1970s. About 70 species of insect pests are recorded on kodo millet.

Shoot and Stem Feeders

The shoot fly *Atherigona simplex* is the primary species attacking kodo millet (Singh and Dias, 1972; Nageshchandra and Ali, 1983; Raghuwanshi and Rawat, 1985). The other occasional species are *A. pulla*, *A. oryzae*, and *A. soccata*. Maggot feeding inside the shoot causes deadheart.

The pink borer *S. inferens* cause damage to the standing crop.

Leaf Feeders

The leaf roller *C. patnalis* and caseworm *Hydrellia philippina* occasionally damage the crop. Armyworms *M. separata* and *S. mauritia* generally occur during the seedling stage of the crop. The skipper butterfly *Pelopidas mathias* is an occasional pest. Thrips, *Stenchaetothrips biformis*, feed on leaves during the seedling stage. The grasshopper *A. exaltata* causes defoliation.

Sucking Pests

The mealy bug *Brevennia rehi* occasionally becomes a serious pest and the affected plants dry up completely (Fig. 1.20). The green leafhopper *Nephotettix nigropictus* sucks the sap from leaves and shoots.

Panicle Pests

The green bug *Nezara viridula*, *Dolycoris indicus*, and the earhead bug *Leptocorisa acuta* are the serious pests on earheads. The nymphs and adults suck the developing grains and cause chaffy grains. Sometimes feeding by the gall midge *Orseolia* spp. causes atrophied ovaries resulting in gall formation.

PROSO MILLET, *PANICUM MILIACEUM*

Proso millet is also called broomcorn millet or common millet. It is generally cultivated in the cooler regions of Asia, eastern Africa, southern Europe, and the United States. Proso millet has adapted well

FIGURE 1.20

Dried up kodo millet plant due to *Brevennia rehi* damage.

Courtesy of A. Kalaisekar.

to temperate plains and high altitudes compared to other millets. The United States is the largest producer of proso millet and produced 250,000 tons during 2006 (USDA-NASS, 2006). The plant looks similar to the wild grass, *Rottboelia cochinchinensis* during the initial stages of growth. The panicles are denser than little millet and from a distance look like panicles of paddy. The ligule with a line of dense hairs is the best identification feature of proso millet. The shallow root system of proso millet makes the crop more efficient in utilizing topsoil moisture and therefore, soil moisture at planting directly affects the grain yield. Proso millet fits well in rotation with annual winter crops such as winter wheat in wheat fallows across the US Great Plains. Crop rotation with proso millet reduces weeds, insect pests, and disease problems in the next crop (IANR, 2008). There are at least 60 species of insects that damage the crop. The important pests are given in the following.

Seedling Pests
The shoot fly *Atherigona pulla* is a major pest in India and Africa. *A. miliaceae*, *A. soccata*, and *A. punctata* are the other shoot fly species recorded on proso millet. Damage results in the formation of deadheart.

The wheat stem maggot *Meromyza americana* occurs in the United States. The maggots feed inside the stem near the upper node and cause whiteheads.

Thrips, *Haplothrips aculeatus*, sometimes cause serious damage to seedlings. Thrips and mites are major seedling pests in the United States. At times the damage by thrips and mites becomes worse and difficult to control.

The armyworms *Mythimna separata*, *M. unipuncta*, *S. exempta*, and *S. frugiperda* cut the seedlings and under dry conditions cause considerable damage.

The field cricket *Brachytrupes* sp. also causes damage similar to that of armyworms.

Stem Borers
Stem borers, *C. partellus*, *Chilo suppressalis*, *Chilo orichalcociliellus*, *S. inferens*, *S. cretica*, and *O. furnacalis*, are the important pests recorded on proso millet. The damage at the vegetative stage produces deadheart and at the panicle stage results in formation of white-ear.

Leaf Feeders
Leaf folders *Cnaphalocrocis medinalis*, and *C. patnalis*; hairy caterpillar *Spilosoma obliqua*; and the rice butterfly *Melanitis leda ismene* are all occasional pests on proso millet.

The Moroccan locust *Dociostaurus maroccanus*, migratory locust *Locusta migratoria*, and grasshoppers *H. banian* and *Oxya chinensis* are recorded as leaf feeders.

Other Pests
The aphid *Sipha flava* is native to North America. Aphids sometimes cause considerable damage in the United States.

Cotton boll worm *Helicoverpa zea* feeds on earhead in the United States. The earhead bug *Leptocorisa acuta* and green bug *N. viridula* suck the milky developing grains in India.

Termites, *Odontotermes* spp. and *Microtermes* spp., are the common species recorded on proso millet during dry seasons in India.

FIGURE 1.21

Larva of *Cnaphalocrocis medinalis* in little millet.

Courtesy of A. Kalaisekar.

LITTLE MILLET, *PANICUM SUMATRENSE*

(Syn. *Panicum miliare*, *Panicum sumatrense* subsp. *psilopodium*)

Little millet, *P. sumatrense*, is native to India and is called Indian millet. The species name is based on a specimen collected from Sumatra (Indonesia) (de Wet et al., 1983). It is mainly cultivated in the Caucasus, China, east Asia, India, and Malaysia. Little millet is adapted to both temperate and tropical climates. It can withstand both drought and water logging. At present the crop is almost restricted to some hilly areas in India and is cultivated on about 500,000 ha. It is an important catch crop in some tribal farms in India. There are comparatively fewer numbers of insect pests reported on little millet, but at least 50 species have been reported. The following are the important insect pests on little millet.

Shoot fly, *A. miliaceae*, causes damage to the seedlings in India (Murthi et al., 1982). Other shoot fly species occurring on little millet are *Atherigona falcata* and *Atherigona lineata*. Deadheart symptoms are produced by their damage.

Armyworms, *M. separata* and *S. frugiperda*, occasionally cause damage under dry conditions during the seedling stage of the crop.

Occasional leaf-feeding caterpillars on foliage are *E. virguncula*, *A. moorei*, *A. albistriga*, and *A. lactinea*.

Flea beetle, *Chaetocnema basalis*, *C. indica*, and *C. denticulata* make shot holes on leaves.

Larvae of the leaf folder *C. medinalis* sometimes occur and feed on leaves by making leaf rolls (Fig. 1.21). Grasshoppers, *Chrotogonus hemipterus*, *Acrida exaltata*, *Aiolopus simulatrix*, and *A. tamulus* generally cause minor leaf damage. The bugs *N. viridula* and *D. indicus* feed on developing grains to cause chaffy or empty earheads. Some species of bugs, like *N. virescens*, *N. nigropictus*, and *Nisia*

FIGURE 1.22

White-ear damage by stem borer in little millet.

Courtesy of A. Kalaisekar.

atrovenosa are reported on little millet. Sugarcane leafhopper, *P. perpusilla*, is a minor pest. Midge, *Orseolia* sp., has been reported on little millet. The spotted stalk borer *C. partellus* infestation at the panicle stage results in production of white-ear (Fig. 1.22). There is one more unidentified lepidopteran borer that also causes considerably high white-ear damage in little millet.

BARNYARD MILLETS, *ECHINOCHLOA ESCULENTA*

(Syn. *Echinochloa utilis, Echinochloa crus-galli* subsp. *utilis*)

E. FRUMENTACEA

(Syn. *Echinochloa crus-galli* subsp. *edulis, Echinochloa crus-galli* var. *frumentaceum*)

There are two species of cultivated barnyard millet, viz., *E. esculenta*, commonly called "Japanese barnyard millet," and *E. frumentacea*, known as billion-dollar grass or "Indian barnyard millet." These two species are often confused with the wild species viz., *E. crus-galli*, known as "barnyard grass," and *Echinochloa colona*, called "jungle rice grass." *E. crus-galli* and *E. colona* are the wild progenitors of *E. esculenta* and *E. frumentacea*, respectively (Hilu, 1994; Sheahan, 2014a,b). Both *E. esculenta* and *E. frumentacea* possess robust plant type and huge compact panicles with awnless grains compared to their wild counterparts. Both species are cultivated in Asia, Africa, and the United States. Barnyard millet can be grown as a reclamation crop to overcome soil salinity. It is also adapted to acidic soil conditions. There are more than 80 species of insect pests recorded on barnyard millet and the important ones are given in the following.

Root-feeding white grubs *Holotrichia* sp., *Anomala dimidiata*, and *Apogonia* sp. are important in India.

Shoot fly *A. falcata* causes considerable damage to the seedlings, and the other species *A. pulla*, *A. simplex, A. soccata, A. oryzae*, and *Atherigona nudiseta* also sometimes occur on barnyard millet in India. Damage causes production of deadheart. The armyworm *M. separata* feeds on seedlings. The

FIGURE 1.23

Aphid infestation in barnyard millet.

Courtesy of A. Kalaisekar.

thrip *Haplothrips ganglbaueri* feeds on leaves of seedlings by scraping the green matter and thereby produces silvery whitish patches on leaves.

Stem borers *S. inferens*, *C. partellus*, and *Chilo diffusilineus* infest the crop in Asia and Africa.

Leafhoppers *Nephotettix cincticeps*, *Sogatella furcifera*, and *Sogatella kolophon*; plant hoppers *Nilaparvata lugens* and *P. maidis*, leaf bug *C. punctiger*, and aphids *H. setariae* and *Macrosiphum eleusines* are important sucking pests (Fig. 1.23).

Grasshoppers *Acrida exaltata*, *Atractomorpha crenulata*, *Hieroglyphus banian*, *H. daganensis*, *H. nigrorepleptus*, *Oxya nitidula* and *Oxya bidentata* feed on the leaves and defoliate. The leaf caterpillar *Euproctis similis* also occasionally is present.

The bugs *Agonoscelis pubescens*, *D. indicus*, and *N. viridula* infest the developing grains.

OTHER MILLETS

Tef, Eragrostis tef

Tef is also known as "Ethiopian millet" owing to its origin. At present, cultivation as a food crop is highly restricted to Ethiopia and some areas in Kenya. Tef is grown as a hay crop in South Africa, Kenya, the United States, and Australia and as a green fodder in India. The grain has an exceptionally good nutritional profile such as balanced essential amino acids (except lysine) and rich in calcium and iron. People who eat tef grains are found to have resistance to hookworm anemia (Winch, 2006).

The central shoot fly *Delia arambourgi* attacks seedlings and causes the central shoots to die.

The Wello-bush cricket *Decticoides brevipennis* is a serious pest feeding on flowers of tef in Ethiopia. The red tef worm *Mentaxya ignicollis* also causes considerable crop loss in Ethiopia.

Tef epilachna beetle *C. similis* is a serious leaf-feeding pest on tef in Ethiopia and Yemen (Beyene et al., 2007). It transmits rice yellow mottle virus in rice (Abo et al., 2001).

Chrysomelid black beetle *Erlangerius niger* is also a major pest; the adults feed on developing grains and leaves of tef.

The stem-boring wasp *Eurytomocharis eragrostidis* is a serious pest on tef and has reduced forage yields by over 70% in the United States (McDaniel and Boe, 1990; Stallknecht et al., 1993).

Browntop Millet, Panicum ramosum
(Syn. *Brachiaria ramosa, Urochloa ramosa*)

Browntop millet as a subsistence crop was cultivated in ancient India (Boivin et al., 2014). At present, it is grown as a grain and forage crop in India (Madella et al., 2013). Grains are used as a boiled whole grain for porridge (Nesbitt, 2005). It is also grown in Africa, Arabia, China, Australia, and the United States (Sheahan, 2014a,b). The insect pests recorded are shoot flies *A. oryzae, A. pulla,* and *A. punctata*; caseworm *Paraponyx stagnalis*; and red hairy caterpillars *A. albistriga* and *A. moorei.*

Fonio, Digitaria exilis *and* Digitaria iburua
Fonio is an indigenous west African crop. There are two cultivated species, viz., *D. exilis*, known as white fonio, and *D. iburua*, black fonio. Fonio is probably the oldest African cereal. White fonio is mainly cultivated from Senegal to Chad and black fonio is grown mainly in Nigeria as well as the northern regions of Togo and Benin. It is the world's fastest maturing cereal and matures in 60–70 days. Grains are rich in the amino acids methionine and cystine (NRC, 1996).

There are some insect pests recorded on fonio: shoot flies *Atherigona* spp., stem borer *C. partellus*, some species of thrips, bugs, and grasshoppers. The white-backed plant hopper *Sogatella furcifera* is the vector of pangola stunt virus.

Guinea Millet, Brachiaria deflexa
(Syn. *Urochloa deflexa, Pseudobrachiaria deflexa*)

Guinea millet is cultivated in the African savanna, especially in Guinea and Sierra Leone. It is also found in east and south Africa. It also occurs in western Asia, Pakistan, and India. It is a semidomesticated weed and there are very few insects, like thrips, bugs, and grasshoppers, that feed primarily on other cereals but take refuge in it during off seasons.

Job's Tears, Coix lacryma-jobi
Job's tears was grown as a food crop by the aboriginal inhabitants of Mongolian origin even before maize was known in south Asia (Arora, 1977). The crop is grown at present in highly isolated places in hilly tracts of south and southeast Asia. Some of the pests recorded on this crop are the stem borers *S. inferens* and *O. furnacalis*. The rice skipper *P. mathias* feeds on its leaves. The thrip *Chaetanaphothrips orchidii*, aphid *R. maidis*, and woolly aphid *Ceratovacuna lanigera* are all found feeding on this plant as an alternate host.

REFERENCES

Abo, M.E., Alegbijo, M.D., Sy, A.A., Misari, S.M., 2001. An overview of the mode of transmission, host plants and methods of detection of rice yellow mottle virus. Journal of Sustainable Agriculture 17, 19–36.

Ananthakrishnan, T.N. (Ed.), 1973. Thrips: Biology and Control. Macmillan, New Delhi, India.

Arora, R.K., 1977. Job's-Tears (Coix lacryma-jobi): a minor food and fodder crop of Northeastern India. Economic Botany 31 (3), 358–366.

Beyene, Y., Hofsvang, T., Azerefegne, F., 2007. Population dynamics of tef epilachna (Chnootriba similis Thunberg) (Coleoptera, Coccinellidae) in Ethiopia. Crop Protection 26, 1634–1643.

Boivin, N., Fuller, D.Q., Korisettar, R., Petraglia, M., 2014. The South Deccan Prehistory Project. Karnatak University, School of Oxford, University College London. http://www.homepages.ucl.ac.uk/~tcrndfu/web_project/home.html.

de Wet, J.M.J., Prasada Rao, K.E., Brink, D.E., 1983. Systematics and domestication of *Panicum sumatrense* (Graminae). Journal d'agriculture traditionnelle et de botanique appliquée 30 (2), 159–168.

Deeming, J.C., 1971. Some species of *Atherigona* Rondani (Diptera, Muscidae) from northern Nigeria, with special reference to those injurious to cereal crops. Bulletin of Entomological Research 62, 133–190.

FAO, 2014. http://faostat3.fao.org/download/Q/QC/E.

Harris, K.M., 1976. The sorghum midge. Annals of Applied Biology 64, 114–118.

Harris, K.M., Nwanze, K.F., 1992. *Busseola fusca* (Fuller), the African Maize Stalk Borer: A Handbook of Information. ICRISAT Information Bulletin, No. 33: vi + 84 pp.

Hilu, K.W., 1994. Evidence from RAPD markers in the evolution of *Echinochloa* millets (Poaceae). Plant Systematics and Evolution 189 (3–4), 247–257.

IANR, 2008. Producing and Marketing Proso Millet in the Great Plains of USA. University of Nebraska. Lincoln, Extension EC 137.

Jagadish, P.S., 1997. Management of insect pests of small millets with special reference to shootfly. In: National Seminar on Small Millets, 23–24, April 1997 TNAU, Coimbatore.

Jagadish, P.S., Neelu, N., 2006. Current status of small millets: shoot fly future research and development requirements. In: Krishnegowda, K.T., Seetharama, N., Khairwal, I.S., Tonapi, V.A., Ravikumar, S., Jayaramegowda, B., Rao, K.V.R. (Eds.), Proceedings of Third National Seminar on Millets Research and Development-Future Policy Options in India (Vol. III: Small Millets), March 11–12, 2004. Organized by All India Coordinated Pearl Millet Improvement Project, Agricultural Research station, Mandor, Jodhpur 342304, Rajasthan, India, in collaboration with Directorate of Millets Development (Department of Agriculture & Cooperation), Mini secretariat, Bani park, Jaipur 302016, Rajasthan, India, National Research Centre for Sorghum & All India Coordinated Sorghum Improvement Project, Rajendranagar, Hyderabad 500030, AP, India and All India Coordinated Small Millets Improvement Project, University of Agricultural Sciences, GKVK campus, Bangalore, India. pp. 55–58.

Jotwani, M.G., Verma, K.K., Young, W.R., 1969. Observations on shoot fly, *Atherigona* spp. Damaging different minor millet. Indian Journal of Otolaryngology 31 (3), 291–293.

Jotwani, M.G., Young, W.R., 1971. Sorghum insect control –here's what's working in India. World Farming 6–11.

Kishore, P., Srivastava, K.P., 1976. Occurrence of cotton grey weevil as a serious pest of sorghum. Entomologists' Newsletter 6 (3), 30–31.

Kundu, G.G., Kishore, P., 1971. New record of parasites of *Sesamia inferens* W. and *Atherigona nudiseta* R. infesting minor millets. Indian Journal of Otolaryngology 33, 466–467.

Madella, M., Lancelotti, C., Garcia-Granero, J.J., 2013. Millet microremains –an alternative approach to understand cultivation and use of critical crops in prehistory. Archaeological and Anthropological Sciences. http://dx.doi.org/10.1007/s12520-013-0130-y.

McDaniel, B., Boe, A., 1990. A new host record for Eurytomocharis eragrostidis Howard (Chalcidoidea: eurytomidae) infesting *Eragrostis tef* in South Dakota. Proceedings of the Entomological Society of Washington 92, 465–470.

Murthi, T.K., Shirole, S.M., Harinarayana, G., 1982. Screening of *Panicum* for shootfly incidence. MILWAI Newsletter 1, 7.

Murthi, T.K., Harinarayana, G., 1989. Insect pests of small millets and their management in India. In: Seetharam, A., Riley, K.W., Harinarayana, G. (Eds.), Small Millets in Global Agriculture. Proc. 1st International Small Millets Workshop, Bangalore, 29th Oct–2nd Nov., 1986. Oxford & IBH, New Delhi. 255–270 pp.

Nageshchandra, B.K., Ali, T.M.M., 1983. Shootfly species on minor millets in Karnataka. MILWAI Newsletter 2, 15.

Nault, L.R., 1983. Origin of leafhopper vectors of maize pathogens in Mesoamerica. In: Proceedings of the International Maize Virus Diseases Colloquium and Workshop. Ohio Agricultural Research and Development Center, Wooster, Ohio, USA, pp. 75–82.

Nesbitt, M., 2005. Grains. In: Prance, G., Nesbitt, M. (Eds.), The Cultural History of Plants. Routledge Press, New York, pp. 45–60.

NRC., 1996. Lost Crops of Africa. Grains Board on Science and Technology for International Development, vol. I. Office of International Affairs, National Research Council, National Academy Press, Washington, DC. 408 p., ISBN:0-309-58615-1.

Nwanze, K.F., Pring, R.J., Sree, P.S., Butler, D.R., Reddy, Y.V.R., Soman, P., 1992. Resistance in sorghum to the shoot fly, *Atherigona soccata*: epicuticular wax and wetness of the central whorl leaf of young seedlings. Annals of Applied Biology 120, 373–382.

Raghuwanshi, R.K., Rawat, R.R., 1985. Chemical control of *Atherigona simplex* Thomas. on Kodo millet with seed and furrow treatments. Indian Journal of Agricultural Sciences 55 (7), 468–470.

Sharma, H.C., Davies, J.C., 1988. Insects and Other Animal Pests of Millets. ICRISAT, Ptencheru, Andhra Pradesh, India. p. 33+28.

Sharma, H.C., Bhatnagar, V.S., Davies, J.C., 1982. Studies on *Mythimna separata* at ICRISAT. Pages 1-44 in Sorghum Entomology Progress Report, 1980/81. International Crops Research Institute for the Semi-Arid Tropics (Limited distribution), Patancheru, A.P. 502 324, India.

Sharma, H.C., Vidyasagar, P., Leuschner, K., 1988. Nochoice cage technique to screen for resistance to sorghum midge (Cecidomyiidae: Diptera). Journal of Economic Entomology 81, 415–422.

Sheahan, C.M., 2014a. Plant Guide for Browntop Millet (*Urochloa ramosa*). USDA-Natural Resources Conservation Service, Cape May Plant Materials Center, Cape May, NJ.

Sheahan, C.M., 2014b. Plant Guide for Japanese Millet (*Echinochloa esculenta*). USDA-Natural Resources Conservation Service, Cape May Plant Materials Center, Cape May, NJ.

Singh, V.S., Dias, C.A.R., 1972. Occurence of different species of *Antherigona* attacking some minor millets at Kanpur. Entomologists' Newsletter 2, 39–40.

Srivastava, A.S., Goel, G.P., 1962. Bionomics and control of red hairy caterpillar (*Amsacta moorei*). Proceedings of the National Academy of Sciences of India, Section B 32 (2), 97–100.

Stallknecht, G.F., Gilbertson, K.M., Eckhoff, J.L., 1993. Teff: food crop for humans and animals. p. 231–234. In: Janick, J., Simon, J.E. (Eds.), New Crops. Wiley, New York.

USDA-NASS, 2006. Crop Production, National Agricultural Statistics Service, Agricultural Statistics Board, U.S. Department of Agriculture.

Waniska, R.D., Rooney, L.W., 2000. Sorghum food and industrial utilization. In: Smith, C.W., Frederiksen, R.A. (Eds.), Sorghum: Origin, History, Technology, and Production. Wiley, New York, pp. 689–729.

Winch, T., 2006. Growing Food: A Guide to Food Production. Springer, 3300 AA Dordrecht, The Netherlands.

SYSTEMATICS AND TAXONOMY

In this chapter the insect pests are grouped based on common names. Systematics and taxonomic aspects of those insects are given in detail. Additionally, distribution and occurrence are discussed.

Any scientific discussion or logical treatment of biological organisms is possible only after systematization has been achieved. Systematics is a focal point of looking at things that may derive or apply to almost any sort of biological study. Taxonomy is "the theory and practice of classifying organisms" (Mayr and Ashlock, 1991). Taxonomy involves classification and nomenclature. "Classification is the ordering of animals into groups (or sets) on the basis of their relationships, that is, of associations by contiguity, similarity, or both." Nomenclature is "the application of distinctive names to each of the groups recognized in any given zoological classification" (Simpson, 1961). Thus, systematics is a broader term that seeks to "study the diversity and kinds of organisms and the relationships"; classification is an essential part of systematics, which in turn forms a subject matter of taxonomy (Mayr and Ashlock, 1991). Basically systematics seeks to infer the evolutionary tree of life or phylogeny, that is, the evolutionary history, of a species or group of related species. Therefore a holistic understanding of any single animal or group of animals requires information on its systematics. We provide the systematics of all the major pests of sorghum in this chapter, which will help in the clear understanding of subsequent chapters.

Before getting into the systematics of insect pests of millets, we would like the reader to know some of the basic aspects of animal taxonomy. Nomenclature is the fundamental aspect of building a proper systematics of biological objects. Therefore, the nomenclatural aspects of plants, animals, and microorganisms are governed by respective international bodies. The International Code of Zoological Nomenclature (ICZN) lays down norms for the naming of all animal taxa.

The 10th edition of *Systema Naturae* was published in the year 1758 and the starting point of zoological nomenclature has been arbitrarily fixed as January 1, 1758. The Rule of Priority as per the ICZN applies to all the taxa described since 1758. All the names of taxa must be in classical Latin or latinized (if in other languages).

For example, *Atherigona soccata* was described as a species by Rondani in the year 1871. Malloch in 1923 described a species as *Atherigona indica*. In the year 1972, Pont examined the *type specimens* of both *A. soccata* and *A. indica* and found that the two were not different species. Two names cannot be assigned to a single taxon. Now, the following questions arise.

Which name should be assigned for such taxon?

On what basis should a particular name of two be assigned?

Is the author who examined both of the type specimens at liberty to assign a different name to such taxon?

Under such circumstances, the Rule of Priority comes into force. As per this rule the earliest name given to a taxon is valid. In our case, therefore, *A. soccata* Rondani, 1871 becomes the valid name and *A. indica* Malloch, 1923 becomes a synonym. The author who examined the two type specimens cannot

Insect Pests of Millets. http://dx.doi.org/10.1016/B978-0-12-804243-4.00002-1

assign a different name without making taxonomic justification. Therefore, the earliest *available name* becomes the *valid name* as per ICZN.

What happens if an author makes proper taxonomic justification and changes the original species combination by shifting a species from one genus to another?

For example, Swinhoe in 1885 described a specimen from Poona (Maharashtra, India) as *Crambus partellus* Swinhoe, 1885. Later on the described species *partellus* was found to match the genus *Chilo* and therefore the original species combination was changed to *Chilo partellus* by Bleszynski and Collins in 1962. Now the valid name is *Chilo partellus* (Swinhoe) Bleszynski and Collins. Here, note the original author Swinhoe is put in parentheses and the author who shifted the original combination is put thereafter outside the parentheses.

What is a type specimen?

The type specimen is the specimen on which the original description is made. There are several kinds of type specimens designated as per the ICZN.

What is available name and what is valid name?

The name assigned to a taxon that is duly published and available to the public is called the available name. An available name assigned to a taxon that satisfies the provisions of the ICZN is called the valid name. All available names are not valid names but all valid names are available names.

In the nomenclature, the names of superfamily, family, subfamily, and tribe end with "oidea," "idae," "inae," and "ini," respectively.

SHOOT FLIES

The shoot flies, belonging to the order Diptera are, by far, the most conspicuous destructive pests in millets. In general, shoot fly species belonging to the families Muscidae and Anthomyiidae are recorded as damage-causing pests in cereals. Several species of the genus *Atherigona*, belonging to Muscidae, are injurious to millets (Jotwani et al., 1969) grown mostly in semiarid tropics of the world. Many species belonging to the genus *Delia* of the Anthomyiidae are serious pests on cereals such as wheat, barley, rye, and oats cultivated in the cooler regions. Nevertheless, some of the species of *Atherigona* and of *Delia* are reported as pests on cereals and millets, respectively.

Muscidae
(Diptera: Brachycera: Muscoidea)
Atherigona **spp.**
(Atherigoninae: Atherigonini: *Atherigona* Rondani, 1856)

The sorghum shoot fly was first reported in India as the *cholam* fly (Fletcher, 1914). The systematic position and taxonomy of *A. soccata* are now fairly clear owing to a comprehensive review by Pont (1972). To make a clear taxonomic understanding of *A. soccata*, we present the relevant information starting from the generic level to all the closely related species of *soccata*.

The genus *Atherigona* was described by Rondani in 1856 with the type species *Anthomyia varia* Meigen, 1826, by original designation. This genus was originally placed under the subfamily Coenosiinae in the family Muscidae (van Emden, 1940). Later, *Atherigona* was transferred to the subfamily Phaoniinae (Hennig, 1965) and currently, the genus is assigned to the Atherigoninae

(Sujatha et al., 2008). Owing to the peculiar head shape, the modified palpi, the reduced thoracic setae, and the male abdomen and trifoliate process, the genus was placed in a separate tribe, Atherigonini (Fan, 1965). The species of *Atherigona* are grouped in two subgenera, *Acritochaeta* Grimshaw and *Atherigona* sensu stricto (s.str.), a division based on characters of adult morphology and biology (Malloch, 1924; Pont, 1972).

There are 276 species of *Atherigona* described globally as of this writing (Pape and Thompson, 2013). Among these species, the following are economically important from the Asian region.

ECONOMICALLY IMPORTANT SPECIES OF *ATHERIGONA* AND THEIR HOST PLANTS

(All the species listed in the following are collected from the hosts mentioned against the species in India and identified by the senior author, A. Kalaisekar.)

Atherigona approximata: Pennisetum typhoides, Sorghum bicolor
Atherigona atripalpis: Setaria italica
Atherigona biseta: S. italica, Setaria viridis
Atherigona falcata: Echinochloa colona, Echinochloa frumentacea, Echinochloa stagnina, Panicum sumatrense
Atherigona miliaceae: Panicum miliaceum, P. sumatrense
Atherigona naqvii: Triticum aestivum, Zea mays
Atherigona oryzae: Oryza sativa, Paspalum scrobiculatum, T. aestivum, Z. mays
Atherigona pulla: P. miliaceum, P. sumatrense, P. scrobiculatum, S. italica
Atherigona punctata: T. aestivum
Atherigona reversura: Cynodon dactylon (turf grasses)
Atherigona simplex: P. scrobiculatum
Atherigona soccata: S. bicolor, Z. mays, Eleusine coracana

SHOOT FLY SPECIES OF MILLETS AND THEIR IDENTIFICATION (PONT, 1972)

Sorghum shoot fly, *A. soccata* Rondani

A. soccata Rondani, 1871: 332 (Fig. 2.1)

Syn. *A. indica* Malloch, 1923: 193; *A. indica* subsp. *infuscata* Emden, 1940: 123; *Atherigona varia* subsp. *soccata* (Rondani, 1871); Hennig, 1961: 502–504; *Atherigona excisa* (Thomson); Avidov, 1961: 296

Sorghum shoot fly, *A. soccata*, is an insect pest of cultivated sorghum in Asia and Africa. In India the species was originally described as *indica* by Malloch (1923). Subsequently Hennig (1965) proposed that all African, Indian, and Mediterranean populations from sorghum belonged to a single species called *varia* subsp. *soccata* Rondani. Pont (1972) convincingly demonstrated *soccata* as a good single species rather than a subspecies. Thereafter, the species has been referred to as *A. soccata*, which was originally described by Rondani in 1871. Thus, the confusion in the nomenclature of this taxon has been cleared and the correct name is *A. soccata*. Probably, this species was the one that was reported as the *jowar* fly and *cholam* fly occurring on sorghum in Tamil Nadu (India) for the first time as a crop pest (Fletcher, 1914).

This species is the most important of the shoot flies and is found throughout Africa and Asia.

FIGURE 2.1

Adult male fly of *Atherigona soccata*.

Courtesy of A. Kalaisekar.

Identification: Species identification is generally done with male adult flies:

- palpi yellow and dark brown interfrontalia
- fore femur entirely yellow
- fore tarsus without erect hairs
- wing with a dark smudge around tip of subcosta
- tergites 1 + 2 without dark spots
- distinct trifoliate process entirely dark brown

A. soccata males are superficially similar to *A. oryzae*, *A. approximata*, and *A. simplex*. The differences are given in Table 2.1.

Rice shoot fly, *A. oryzae* Malloch

A. oryzae Malloch, 1925: 117

Syn. *Atherigona exigua* Stein, Dammerman, 1919: 83; *Atherigona seticauda* Malloch, Cendana and Calora, 1967: 592; *A. excisa* (Thomson); Meksongsee, Prachaubmoh and Sepsawadi, 1968: 532; *A. indica* Malloch; Grist and Lever, 1969: 258

A. oryzae attacks rice in the Asian and Australia regions (Pont, 1972). It is also recorded on some millets:

- males with yellow palpi and dark brown interfrontalia
- fore femur apically darkened
- wing with a dark smudge around tip of subcosta
- fore tibia and tarsus dark; tibia basally yellow

Table 2.1 Distinguishing Characters of Superficially Similar Species of *Atherigona*

Character	*soccata*	*oryzae*	*approximata*	*simplex*
Palpi	Yellow	Yellow	Yellow	Dark
Interfrontalia	Dark brown	Dark brown	Yellow	Black
Fore femur	Entirely yellow	Apically dark	Entirely yellow	Apically dark
Fore tibia	Yellow, dark at tip	Apically dark	Yellow, dark at tip	Apically brown
Fore tarsus	Yellow	Dark	Yellow	Brown
Hypopygial prominence	Less prominent	Less prominent	Distinct	Distinct
Trifoliate process	Distinct, entirely dark brown	Distinct, thickening on stalk	Median piece end fork shaped	Median piece slightly sinuate

- stripes on mesonotum indistinct
- hypopygial prominence less distinct
- trifoliate process with thickening on stalk
- median of trifoliate process totally membranous

Pearl millet shoot fly, *A. approximata* Malloch

A. approximata Malloch, 1925; 118

> **Syn.** *Atherigona* sp. near *approximata* Malloch; Jotwani, 1969: 293; Pradhan, 1971: 103
> This species was reported as the *cumbu* (= pearl millet) fly (Fletcher, 1917; Moiz and Naqvi, 1968) on *Pennisetum typhoides* and also on *Sorghum bicolor*. The species is distributed in south and western India.
> This species can be identified by the following characters:

- palpi, interfrontalia, and fore femur entirely yellow
- tarsal segments 2–5 with erect anterodorsal hairs
- no dark smudge at tip of subcosta
- fore tibia yellow, darkening at tip
- abdominal tergites 1 + 2 and 5 lack dark spots
- female tergite 8 with straight anterior margin

Foxtail millet shoot fly, *A. atripalpis* Malloch

A. atripalpis Malloch, 1925: 116

> This species was recorded on *Setaria italica*, *S. glauca*, and *S. plicata* from China and south Asia:

- palpi and interfrontalia entirely dark brown
- fore femur partly darkened, apex of subcosta lacks a dark smudge
- fore tarsus without erect hairs
- abdominal tergite 5 lacks dark spots
- moderate to indistinct stripes on mesonotum
- tridentate hypopygial prominence

A. biseta Karl

A. biseta Karl, 1939: 279–280

> This is said to be a specific pest of foxtail millet in China and running only two generations in a year (Wenn, 1964). The host range is restricted to a very few *Setaria* species, namely, *S. viridis*, *Setaria faberi*, and *Setaria pumila*.

A. biseta possesses a distinctly tridentate hypopygial prominence and a trifoliate process with strongly enlarged apically excised median piece. In the closely similar species *A. atripalpis*, the tip of the median piece of the trifoliate process is not excised (Pont and Magpayo, 1995).

Barnyard millet shoot fly, *A. falcata* (Thomson)

A. falcata (Thomson) Stein, 1910: 77; original species combination *Coenosia falcata* Thomson, 1869: 560

Syn. *Atherigona nudiseta* Malloch, 1923: 186; *Atherigona quadripunctata* (Rossi); Hennig, 1941: 208; *A. (Atherigona) nudiseta nudiseta* Malloch; Fan, 1965: 69; *A. (Atherigona) nudiseta megaloba* Fan, 1965: 69

This species is distributed throughout the Asian region and *E. frumentacea* was recorded as its main host. The species is comparatively larger in size than other species:

- palpi, interfrontalia, and fore femur entirely yellow
- subcosta without a dark apical smudge
- fore tarsus with erect anterodorsal hairs on segments 2 or 3–5
- fore tibia and tarsus yellow
- inconspicuous stripes on mesonotum
- prominent hypopygium and knoblike and quite easily recognized
- trifoliate process with broad and flattened lateral plates; narrow and small median piece

Kodo millet shoot fly, *A. simplex* (Thomson)

A. simplex (Thomson) Stein, 1910: 77; original combination: *Coenosia simplex* Thomson, 1869: 560

Syn. *Atherigona bituberculata* Malloch, 1925: 119; *A. (Atherigona) bituberculata* Malloch; Fan, 1965: 70

This species was recorded on kodo millet and is distributed widely:

- palpi entirely dark
- shining black parafrontalia
- hypopygial prominence, like that of *approximata*
- trifoliate process with slightly sinuate median piece

Proso millet shoot fly, *A. pulla* (Wiedemann, 1830)

Original species combination: *Coenosia pulla* Wiedemann, 1830: 441

Syn. *Atherigona destructor* Malloch, 1923: 185; *A. (Atherigona) destructor* Malloch; Fan, 1965: 68

It was recorded on *Panicum* spp. and is distributed in south India. This species can be easily recognized by the yellow vibrissal bristles in both sexes:

- interfrontalia, palpi, and fore femur entirely yellow
- fore tibia yellow, brown at tip
- fore tarsus with long erect anterodorsal hairs on segments 1–5
- no dark smudge at tip of subcosta
- hypopygium less prominent and knoblike
- trifoliate process with short stalk and median piece thickened at base

Little millet shoot fly, *A. miliaceae* Malloch

A. miliaceae Malloch, 1925: 118

This species was reported from India and China on *Panicum* spp.:

- interfrontalia, palpi, and fore femur yellow
- subcosta without a dark smudge at tip
- knoblike hypopygial prominence

***A. punctata* Karl**

A. punctata Karl, 1940: 147; Hennig, 1941: 208

Coimbatore wheat stem fly; Ramachandra Rao, 1924: 334:

- very close to *A. miliaceae* Malloch
- yellow palpi and interfrontalia
- differs from *A. miliaceae* and *A. punctata* in having smaller hypopygial prominence
- median piece of trifoliate process produced in an arc beyond the lateral apical setae

***A. varia* (Meigen)**

A. varia (Meigen) Seguy, 1937: 226; *A. varia* Meigen, 1826: 187 (original combination)

Syn. *A. (Atherigona) quadripunctata* (Rossi); Fan, 1965: 68

This species is most closely related to *ferruginea* Emden, whereas *soccata* is closer to *oryzae*. The hosts of *varia* are completely unknown (Pont, 1972).

Wheat stem fly, *A. naqvii* Steyskal

A. naqvii Steyskal, 1966: 53

This species is widespread and recorded on *T. aestivum:*

- trifoliate process with two long pale setae at tip of median piece
- hypopygium less prominent and knoblike with three low tubercles
- palpi and interfrontalia yellow
- fore femur partly darkened
- wing lacks a dark spot at the tip of the subcosta

Tomato fly, *Atherigona orientalis* Schnier

A. orientalis, Schnier, 1868: 295

Syn. *Coenosia excisa* Thomson, 1869: 560; *Atherigona trilineata* Stein, 1900: 157; *Acritochaeta pulvinata* Grimshaw, 1901; 42; *A. (Acritochaeta) excisa* (Thomson); Stein, 1910: 76; *A. excisa* (Thomson); Stein, 1915: 42; *A. exigua* Stein; Meijere, 1918: 357; *A. excisa* (Thomson); Malloch, 1923: 185; *A. excisa* var. *flavipalpis* Malloch, 1928: 303; *Atherigona flavipalpis* Malloch; Synder, 1965: 248

This species belongs to the subgenus *Acritochaeta*. Larvae are saprophagous and develop in the decaying plant tissues that are already infested by shoot flies and stem borers. It is sometimes found in already infested sorghum and maize:

- complete absence of hypopygial prominence as well as trifoliate process
- males with preapical dorsal excavation on fore femur
- elongate palpi

Anthomyiidae

(Diptera: Brachycera: Muscoidea)

Barley shoot flies, *Delia arambourgi* (Seguy)

Syn. *Phorbia arambourgi* (Seguy, 1938); original species combination: *Hylemya arambourgi* Seguy, 1938
Delia flavibasis **(Stein)**

Original species combination: *Hylemya flavibasis* Stein, 1903

These two species of *Delia* are recorded on tef in Ethiopia, Kenya, and Tanzania. These flies are primarily pests of barley. Other hosts that support *D. arambourgi* and *D. flavibasis* are maize, wheat, pearl millet, and some grasses.

BORERS AND CATERPILLARS

The lepidopteran pests causing minor to devastating economic losses to the millets are arbitrarily grouped into borers and caterpillars under this subheading. It is to be noted that the borers that are described in the following are also caterpillars.

BORERS

The stem borers infesting millets that we deal in this book belong to two superfamilies under the order Lepidoptera.

1. Superfamily: Pyraloidea
   a. Family: Crambidae
      i. Subfamily: Crambinae
         Chilo partellus (Swinhoe)
         C. auricilius Dudgeon
         C. infuscatellus Snellen
         C. sacchariphagus (Bojer)
      ii. Subfamily: Pyraustinae
         Ostrinia furnacalis Guenée
   b. Family: Pyralidae
      i. Subfamily: Phycitinae
         Maliarpha separatella Rogonot
2. Superfamily: Noctuoidea
   a. Family: Noctuidae
         Sesamia inferens (Walker)

Pyraloidea is the third largest superfamily of the Lepidoptera after Noctuoidea and Geometroidea. The group includes about 16,000 species worldwide, with its greatest richness in the tropics. Morphologically, the superfamily is defined by a basally scaled proboscis and the presence of abdominal tympanal organs (Solis, 2007).

Pyraloidea currently is divided into two families based primarily on two distinct tympanal organ types of the adult abdomen, but characters of the larvae also confirm this division. The Pyralidae have a tympanal case that is almost closed, the conjunctiva and tympanum are in the same plane, and the praecinctorium (a structure that joins two tympanic membranes) is absent. The Crambidae have a tympanal case that is open with a wide anteromedial aperture, the conjunctiva and tympanum are in a different plane and meet at a distinct angle, and the praecinctorium is present (Minet, 1981; Maes, 1995). Börner (1925) was the first to recognize the difference between the two groups in the Pyraloidea, and Munroe (1972, 1973, 1976) proposed the informal groups Pyraliformes and Crambiformes based on the major differences between the two types of tympanal organs. Subsequently, Munroe's groups were elevated to the Pyralidae and Crambidae based on an extensive study of tympanal organs in Lepidoptera (Minet, 1983; Maes, 1998).

There has been confusion in assigning family and subfamily to the genus *Chilo* belonging to the superfamily Pyraloidea. Many reports wrongly refer the genus *Chilo* to the subfamily Crambinae under the family Pyralidae.

Crambidae: The tympanum and conjunctivum make a clear angle and do not lie along the same plane (subfamilies: Crambinae and Schoenobiinae).

Pyralidae: The tympanum and conjunctivum do not make a clear angle and do lie along the same plane (subfamilies: Phycitinae and Galleriinae).

Therefore, the genus *Chilo* undoubtedly lies under the family Crambidae and subfamily Crambinae, as it satisfies the earlier classification criteria.

The identification marks of borers belonging to *Chilo* in the following text are based on Bleszynski (1970).

Family: Crambidae

Subfamily: Crambinae Latreille, 1810

Tribe: Chiloini Heinemann, 1865

Genus: *Chilo* Zincken, 1817

Chilo species originated in Asia and spread to Africa.

Spotted stem borer, *Ch. partellus* (Swinhoe, 1885) (Fig. 2.2)

FIGURE 2.2

Adult moth of *Chilo partellus*.

Courtesy of A. Kalaisekar.

Original species combination: *Crambus partellus* Swinhoe, 1885: 875

Syn. *Chilo simplex* Butler, 1880: 690; *Crambus zonellus* Swinhoe, 1884: 528; *Chilo lutulentalis* Tams, 1932: 127; *Ch. partellus acutus* Bhattacherjee, 1971: 298; *Chilo partellus coimbatorensis* Bhattacherjee, 1971: 299–301; *Chilo partellus kanpurensis* Bhattacherjee, 1971: 302

Chilo partellus was first described by Swinhoe (1885) as *Crambus partellus*. The spotted stem borer is native to Asia and is a pest of sugarcane, maize, and sorghum. In India, the earliest record of *C. partellus* was made by E.C. Cotes (1889) as the sorghum borer and it was exceedingly abundant in the Central Provinces (Indian Museum Notes, 1900). Many species of *Chilo* develop on cereal crops and wild grasses, mostly in the tropics. A thorough revision of all known species of *Chilo* in the world was published by Bleszynski (1970), and in it, *Chilo zonellus* (Swinhoe), a widely used name in the literature, was made a synonym of *C. partellus*. At present, therefore *C. partellus* is the valid name for the spotted stem borer. This species is distributed in Asia and later spread to Africa.

It is a serious pest of sorghum and maize. It is also recorded on other millets.

Male:

- body yellowish brown, suffused with fuscous
- ocellus well developed
- face distinctly conical, with distinct corneous point and ventral ridge slight (Fig. 2.2)
- forewing with costal area darkish
  - *R1* free
  - traces of dark specks below middle of cell and at lower angle
  - the veins of outer area slightly streaked with fuscous
  - a marginal series of black specks (Fig. 2.2)
  - discal dot present (Fig. 2.2)
  - metallic scales absent
- hind wing dirty white to gray with slight fuscous tinge

Female: body and forewing paler; hind wing white

Sugarcane stalk borer, *C. auricilius* Dudgeon, 1905: 405

Syn. *Diatraea auricilia* (Dudgeon): Fletcher 1928: 58; Gupta 1940: 799; *Chilotraea auricilia* (Dudgeon): Kapur 1950: 408; *Chilo popescugorji* Bleszynski 1963: 179; *Chilo auricilia* Dudgeon: Bleszynski and Collins 1962: 239; *C. auricilius* Dudgeon; Bleszynski 1965: 113; 1969: 16

The species is distributed in India, Nepal, Bangladesh, Myanmar, Sri Lanka, China, Malaysia, Indonesia, Papua New Guinea, Philippines, Taiwan, Thailand, and Vietnam. The hosts of economic importance are sugarcane, sorghum, maize, and rice.

Identification:

- ocellus small and distinct
- face produced forward, smooth, or with small point and ventral ridge absent
- forewing
  - *R1* confluent with *Sc*
  - ground-color yellow or brownish
  - variably irrorated with brown scales

- discal dot present
- terminal dots large and fringe shiny golden
- subterminal line close to termen with row of metallic scales
- median line diffused with subterminal line
- few small silvery specks in middle of wing
- hind wing brownish

Sugarcane early shoot borer, *C. infuscatellus* Snellen, 1890: 94

Syn. *Argyria coniorta* Hampson, 1919: 449; *Argyria sticticraspis* Hampson, 1919: 449; *Chilo tadzhikiellus* Gerasimov, 1949: 704; *Diatraea calamina* Hampson, 1919: 544; *Diatraea shariinensis* Eguchi, 1933: 3

The species is distributed in south and southeast Asia. It is primarily a pest of sugarcane. It also infests maize, sorghum, and rice.

Identification:

- ocellus well developed
- face rounded, slightly protruding forward beyond eye
- forewing
  - length 10.0–13.0 mm
  - *R1* confluent with *Sc*
  - ground-color and maculation very variable, dull, from light sand-yellow to chocolate-brown
  - discal dot present or variably reduced
  - transverse lines present or absent
  - terminal dots present
  - metallic scales absent
- Hind wing dirty white in male and silky white in female

Sugarcane internode borer, *C. sacchariphagus* (Bojer, 1856)

Original species combination: *Procera sacchariphagus* Bojer, 1856

Syn. *Borer saccharellus* Guenée, 1862; *Chilo mauriciellus* Walker, 1863; *Diatraea striatalis* Snellen, 1891; *Chilo venosatus* Walker, 1863; *Argyria stramineella* Caradja, 1926; *P. sacchariphagus indicus* Kapur, 1950; *Argyria sacchariphagus stramineella* Caradja, 1926

Three subspecies are recognized in this species based on genitalia differences due to geographical locations.

C. sacchariphagus sacchariphagus has been recorded in Malaysia, Indonesia, and the Indian Ocean islands.

C. sacchariphagus stramineellus is reported from the South of China and Taiwan.

C. sacchariphagus indicus has been found in India.

It is distributed in Asia and the Indian Ocean islands. This is also an important pest of sugarcane and attacks sorghum and maize.

Identification:

- ocellus reduced
- face rounded, not protruding forward beyond eye
- corneous point and ventral ridge both absent

- labial palpus three (male) to four (female) times as long as diameter of eye
- Forewing
  - *R1* confluent with *Sc*
  - length 12.0–18.0 mm, maximum width 4.5–6.0 mm
  - apex acute; ground-color dull light brown
  - veins and interneural spaces outlined with whitish beige
  - discal dot distinct, often double
  - terminal dots present
  - transverse lines absent
  - fringes slightly glossy, concolorous or lighter than the ground-color
- Hind wing dirty white to light brown in male, silky whitish in female

Subfamily: Pyraustinae Meyrick, 1890

Genus: *Ostrinia* Hübner, 1825

Asian corn borer, *O. furnacalis* Guenée, 1854

Syn. *Botys damoalis* Walker, 1859: 656; *Botys salentialis* Snellen, 1880: 207; *Pyrausta polygoni* Dyar, 1905: 955; *Pyrausta vastatrix* Schultze, 1908: 35; *Spilodes kodzukalis* Matsumura, 1897: 237

This species is distributed in south and southeast Asia, the Indian Ocean islands, and Australia. It is a polyphagous species and causes yield losses in maize and sorghum. Eggs are laid in masses of 5–50. Larvae are light brown dorsally with three longitudinally running series of dark spots; matured larva is about 2–3 cm; pupa is reddish brown, 2–2.5 cm long.

Identification:

- adult moths with forewing
  - tawny brown or brownish yellow
  - brown wavy lines distally more distinct
  - dark discal spot
  - hind wing
  - dull creamy brown without any markings

The other important species is ***Ostrinia nubilalis* (Hübner, 1796)**.

Millet stem borer, *Coniesta ignefusalis* (Hampson, 1919)

Syn. *Diatraea ignefusalis* Hampson, 1919; *Chilo pyrocaustalis* Hampson, 1919

This is a persistent pest on pearl millet in Africa (Youm et al., 1996).

Corn borer, *Diatraea grandiosella* Dyar, 1911

This infests maize and sorghum in America and pearl millet in Africa.

Family: Pyralidae
Subfamily: Phycitinae Zeller, 1839
Green-striped borer, *Maliarpha separatella* (Rogonot, 1888)
(Genus: *Maliarpha* Rogonot, 1888) (De Prins and De Prins, 2013)

Syn. *Anerastia pallidicosta* **Hampson, 1896;** *Enosima vectiferella* **Ragonot, 1901**

This species is widely distributed in Africa. It is also reported from Asia, especially India, Myanmar, and China. It is a pest of rice, maize, and sorghum.

Identification:

- wing span 40–50 mm.
- pale yellow or brownish forewings
- conspicuous broad longitudinal reddish brown band runs below costal margin from wing base to apex
- both wings without vein M3
- hind wings white with metallic glossiness and fringed with long hairs

White borer, *Saluria inficita* Walker, 1863

This infests maize and sorghum in Asia and Africa.

Subfamily: Gallerinae

African sugarcane stalk borer, *Eldana saccharina* Walker, 1865

Syn. *Eldana conipyga* Strand, 1912; *Ancylosidia conipyga* Strand, 1913

This is primarily a pest of sugarcane indigenous to Africa and is distributed in sub-Saharan Africa. It damages sorghum and rice.

Family: Noctuidae

Subfamily: Hadeninae Guenée, 1837

Genus: *Sesamia* Guenée, 1852

Sesamia is one of the most multitudinous genera, with 157 described species (Moyal, 2006). Most of the species of these noctuid borers spread from Africa to Europe, Central Asia, and Asia (Moyal et al., 2011). These species are African in origin.

Cereal pink borer, *S. inferens* (Walker, 1856) (Fig. 2.3)

Original species combination: *Leucania inferens* Walker, 1856, 9: 105

Syn. *Leucania proscripta* Walker, 1856, 9: 106; *Sesamia tranquilaris* Butler, 1880, 1880: 674; *Nonagria gracilis* Butler, 1880: 675; *Sesamia albiciliata* Snellen, 1880, 23: 44; *Nonagria innocens* Butler, 1881: 173; *Sesamia creticoides* Strand, 1920; *Sesamia kosempoana* Strand, 1920; *Sesamia sokutsuana* Strand, 1920; *Sesamia hirayamae* Matsumura, 1929

Pink stem borer is polyphagous and distributed in the Asian and the Palearctic region.

The adult moth resembles *Mythimna* species and differs in having smooth hairy eyes; the forewing is much less striate, central streak diffusely darker, and margin darker brown.

Identification:

- moth with pale yellow brown body
- a thick tuft of hair on head and thorax (Fig. 2.3)
- forewings light brown with brown spots throughout the wings
  - apically radiating faint purplish red band bordered by light stripes
- hind wing whitish with light yellow lines along major veins

Species complexes, namely, *grisescens* Warren (New Guinea, Seram), *arfaki* Bethune-Baker (New Guinea) in tropical Australasia, *uniformis* Dudgeon in India and China, and a diverse African species complex are identified based on male genitalia (Wu, 1981).

Other important species, namely, **Sesamia calamistis Hampson, 1910**, and **Sesamia cretica Lederer, 1857**, are present in Africa.

African maize stalk borer, *Busseola fusca* Fuller, 1901

Syn. *Busseola sorghicida* Thurau; *Calamistis fusca* Fuller, 1901

FIGURE 2.3

Adult moth of *Sesamia inferens*.

Courtesy of A. Kalaisekar.

Detailed descriptions of this species (Hampson, 1901) show that it is closely related to the widespread genus *Sesamia*. This group of species found in Africa is thoroughly revised and described (Tams and Bowden, 1953). It is distributed throughout sub-Saharan Africa and is a serious pest on millets in African dry savanna regions (Kfir et al., 2002).

CATERPILLARS

Several species of lepidopteran caterpillars belonging to five families, namely Crambidae, Erebidae, Hesperiidae, Noctuidae, and Pyralidae, are recorded as pests on millets. The majority of the caterpillars are known to cause damage that usually warrants no intervention. However, some of them sporadically cause severe damage to the crops, especially armyworms and cutworms, on sorghum in India and in Africa, and on tef in Africa.

Family: Crambidae
Leaf folders, *Cnaphalocrocis* Lederer, 1863: 384

Syn. *Marasmia* Lederer, 1863: 385; *Dolichosticha* Meyrick, 1884: 304; *Marasmia* Hampson, 1898: 638; *Epimima* Meyrick, 1886: 235; *Lasiacme* Warren, 1896: 176; *Bradinomorpha* Matsumura, 1920: 514; *Susumia* Marumo, 1930: 41; *Prodotaula* Meyrick, 1934: 541; *Neomarasmia* Kalra, David and Banerji, 1967: 544

Rice leaf folder, *Cnaphalocrocis medinalis* (Guenée, 1854)

Syn. *Salbia medinalis* Guenée 1854; *Botys iolealis* Walker 1859; *Botys rutilalis* Walker 1859; *Cnaphalocrocis jolinalis* Lederer 1863; *Botys acerrimalis* Walker 1865

The discovery of a sympatric species, *Cnaphalocrocis patnalis* (Bradley, 1981), which is morphologically similar and often indistinguishable from *C. medinalis*, exposed the general attitude of casually handling the species identity of insect pests. Thus ambiguity arises as to whether all the research data generated before 1981 really belonged to *C. medinalis*!

The other important species of leaf folders is ***C. patnalis* (Bradley, 1981)**. Syn. *Cnaphalocrocis trapezalis* (Guenée, 1854); *Marasmia trapezalis* Guenée, 1854; original species combination: *Marasmia patnalis* Bradley, 1981.

Family: Erebidae Leach, 1815
Subfamily: Arctiinae Leach, 1815

Red hairy caterpillars cause sporadic damage. The caterpillars feed gregariously on leaves; initial instars cause leaf skeletonization and later instars defoliate the complete leaf. The important species are given next with differentiating characters (as given by Hampson, 1901).

***Amsacta albistriga* (Walker, 1864)**. Original species combination: *Aloa albistriga* Wlk., 1864: 303. Its head and thorax are white. Palpi are yellow; third joint is black, antenna black. Back of the head, stripes on shoulders, fore coxae, and femora above are yellow. Legs are striped with black, tarsi ringed with black. Abdomen is orange yellow. Forewing is pale brown, costa are yellow, two discoidal black point. Hind wing is pure white, black discoidal lunule. It is distributed in south India. It has been recorded on finger millet and sorghum.

***Amsacta lactinea* (Cramer, 1777)**. Original species combination: *Bombyx lactinea* Cram., 1777: 133; Syn. *Bombyx sanguinolenta* F., 1793:473; *Aloa marginata* Moore, 1883: 15; *Rhodogastria frederici* Kirby, 1892: 223. Its palpi are crimson, black at tips. Antenna is black. Vertex of the head, neck, edges of tegulae, stripes on shoulders, fore coxae with black patches, fore and middle tibiae are striped with black, tarsi are banded with black. Forewing is with crimson costal fascia, black points usually at angles of the cell. Hind wing is with discoidal black lunule and four subterminal spots, any of which may be absent. These species are recorded on finger millet.

***Amsacta moorei* (Butler, 1876)**. Original species combination: *Areas moorei* Butl., 1876: 23; Syn. *Aloa sara* Swinh., 1889: 404. Moth is white, palpi scarlet at sides and at tip, and antenna is black, scarlet at back and basal joint of head. It has streaks on its shoulders, fore coxae at sides, and femora above scarlet, the fore coxae are with black spots, tibiae streaked, and tarsi ringed with black, abdomen is scarlet. Forewing is with scarlet costal fascia, black point at angles of the cell. Hind wing is with discoidal black lunule and three almost terminal spots. This species is distributed widely in the world. It has been recorded on finger millet, sorghum, and pearl millet.

***Barsine roseororatus* (Butler, 1877)** (Fig. 2.4). Original species combination: *Ammatho roseororatus* Butler, 1877: 341. It has been recorded on millets in India.

***Creatonotus gangis* (Linnaeus, 1763)**. Original species combination: *Phalaena gangis* Linnaeus, 1763. Syn. *Creatonotos continuatus* Moore, 1877; *Noctua interrupta* Linnaeus, 1767; *Creatonotos flavoabdominalis* Bang-Haas, 1938. It has been recorded on millets in India. Another common species is ***Creatonotus transiens* (Walker, 1855)**. Original species combination: *Spilosoma transiens* Walker, 1855: 675.

***Conogethes punctiferalis* (Guenée, 1854)**. Original species combination: *Astura punctiferalis* Guenée, 1854. Syn. *Deiopeia detracta* Walker, 1859; *Botys nicippealis* Walker, 1859; *Astura guttatalis* Walker, 1866. This species feeds on earheads of sorghum in Asia.

***Cyana horsfieldi* Roepke, 1946** (Fig. 2.5). It has been recorded on sorghum in India.

***Spilosoma obliqua* Walker, 1855**. Syn. *Arctia montana* Guérin-Méneville, 1843; *Spilosoma suffusa* Walker, 1855; *Spilosoma dentilinea* Moore, 1872; *Spilarctia confusa* Butler, 1875; *Diacrisia assamensis* Rothschild, 1910. This species is called Bihar hairy caterpillar in India. It is a common defoliator on sorghum and other millets in India.

FIGURE 2.4

Adult moth of *Barsine roseororatus*.

Courtesy of A. Kalaisekar.

FIGURE 2.5

Adult moth of *Cyana horsfieldi*.

Courtesy of A. Kalaisekar.

Subfamily: Lymantriinae Hampson, 1893

Hairy caterpillars

***Euproctis limbalis* (Herrich-Schäffer, 1855)**. Original species combination: *Porthesia limbalis* Herrich-Schäffer,1855. Syn. *Urocoma boeckeae* Herrich-Schäffer,1858; *Ela leucophaea* Walker, 1862; *Nygmia limbalis* Swinhoe, 1923. These caterpillars are found feeding on leaves and earhead in sorghum and other millets.

Yellow-tail moth, *Euproctis similis* (Fuessly, 1775) (Fig. 2.6). Original species combination: *Phalaena similis* Fuessly, 1775. This moth feeds on millets in the Asian region.

***Euproctis virguncula* Walker, 1855**. Syn. *Euproctis marginalis* Walker, 1856: 1731; *E. virguncula javanica* Strand, 1915: 333; *Arctornis virguncula*; Swinhoe, 1922: 477; *Porthesia virguncula* Collenette, 1932: 58.

Tussock moths

***Arna micronides* Eecke van, 1928** (Fig. 2.7, female moth). This moth feeds on millets in India.

FIGURE 2.6

Adult moth of *Euproctis similis*.

Courtesy of A. Kalaisekar.

FIGURE 2.7

Adult female moth of *Arna micronides*.

Courtesy of A. Kalaisekar.

Nygmia amplior **(Collenette, 1930)** (Fig. 2.8, male moth). Original species combination: *Euproctis amplior* Collenette, 1930; *Nygmia plana* **Walker, 1856**; *Nygmia guttulata* **(Snellen, 1886)**. Original species combination: *Euproctis guttulata* Snellen, 1886 and *Nygmia barbera* **(Swinhoe, 1903)** (Fig. 2.9, female moth). Original species combination: *Euproctis barbera* Swinhoe, 1903: 197. Caterpillars of these moths feed on millets in India.

Orvasca subnotata **Walker, 1865: 502** (Fig. 2.10, female moth). There is another species, which is found in the Indian subregion, *Orvasca limbata* **Butler, 1881**. These moths feed on millets in the Asian region.

Family: Hesperiidae

Rice skipper butterfly, *Pelopidas mathias* Fabricius, 1798

The species was described from a specimen collected from India. A 2005 reexamination of the type specimen seeks to shift the species to *Borbo* (Larsen, 2005). There are three subspecies recognized (Evans, 1949), namely, *P. mathias oberthüri* in the Asia Pacific, *P. mathias repetita* in Papua New Guinea, and *P. mathias mathias* in southeast and south Asia and Africa.

FIGURE 2.8

Adult male moth of *Nygmia amplior*.

Courtesy of A. Kalaisekar.

FIGURE 2.9

Adult female moth of *Nygmia barbera*.

Courtesy of A. Kalaisekar.

FIGURE 2.10

Adult female moth of *Orvasca subnotata*.

Courtesy of A. Kalaisekar.

This species feeds on leaves of finger millet, sorghum, and kodo millet in Africa and Asia.

Family: Noctuidae

Cutworm

Caterpillars feed on seedlings during the night and they cut the seedlings at ground level, giving an appearance of grazing.

Black cutworm, *Agrotis ipsilon* **(Hufnagel, 1766)**. Original species combination: *Phalaena ipsilon* Hufnagel, 1766. **Syn.** *Noctua suffusa* Denis and Schiffermüller, 1775; *Noctua ypsilon* Rottemburg, 1777; *Phalaena idonea* Cramer, 1780; *Bombyx spinula* Esper, 1786; *Agrotis telifera* Harris, 1841; *Agrotis bipars* Walker, 1857; *Agrotis frivola* Wallengren, 1860; *Agrotis aneituna* Walker, 1865; *Agrotis pepoli* Bertolini, 1874; *Agrotis aureolum* Schaus, 1898.

This species is widespread and polyphagous. It has the ability to migrate thousands of miles.

Armyworms

Armyworms also feed during the night and completely defoliate the leaves, leaving only the midrib. Sometimes armyworms damage the earhead also. They are recorded on pearl millet, finger millet, sorghum, and foxtail millet. Armyworms are sporadic pests and sometimes assume outbreaking proportions. The two important genera that contain the most serious armyworms of the world are *Mythimna* and *Spodoptera* (Sutrisno, 2012). Species of *Mythimna* are more specific feeders of Graminae while the *Spodoptera* species are polyphagy without any specificity of hosts plants (Sutrisno, 2012). Some of the important species are the following.

Subfamily: Hadininae

Mythimna **spp.**

The *Mythimna* complex has a typical "dead grass" appearance of the forewing. The hind wing lacks *M2* and only three veins branch off from the posterior sector of the discal cell. An oblique, bladelike ovipositor in the female and a small pouch between the bases of the valves of the male genitalia are unique characters of *Mythimna* (Holloway, 1989). Seven subgenera are designated under *Mythimna* based on genitalia of both male and female: *Acantholeucania, Anapoma, Dysaletia, Hephilare, Mythimna, Pseudaletia,* and *Sablia* (Yoshimatsu,1994). Under the genus *Mythimna*. Identification keys are available for this genus (Barrion and Litsinger, 1994).

Mythimna separata (Walker, 1865) Original species combination: *Leucania separata* Walker, 1865. Syn. *Leucania consimilis* Moore, 1881; *Pseudaletia separata* Franclemont, 1951. This species belongs to subgenus *Pseudaletia.* It is an important species damaging millets.

***M. loreyi* (Duponchel, 1827)** Original species combination: *Noctua loreyi* Duponchel, 1827; Syn. *Noctua caricis* Treitschke, 1835; *Leucania collecta* Walker, 1856; *Leucania exterior* Walker, 1856; *Leucania denotata* Walker, 1856; *Leucania loreyi* Duponchel, 1827. This species belongs to subgenus *Acantholeucania.* Caterpillars of this species also damage millets.

Subfamily: Noctuinae

Spodoptera **spp.**

The genus *Spodoptera* contains 22 species; ***Spodoptera exigua* (Hübner),** *Spodoptera littoralis* **(Biosduval),** and ***Spodoptera cilium* Guenée** in Africa; ***Spodoptera litura* (Fabricius)** in India, the Far East, and Australasia; ***Spodoptera frugiperda* (Smith)** in North America and the Caribbean; and ***Spodoptera mauritia* (Biosduval)** in Asia are important species. There are two subspecies recognized under *S. mauritia*, namely, *S. mauritia mauritia*, restricted to Africa, and *S. mauritia acronyctoides*, found in the Asian region (Fletcher, 1956).

African armyworm, *Spodoptera exempta* (Walker, 1856). Original species combination: *Agrotis exempta* Walker, 1856. Syn. *Prodenia bipars* Walker, 1857; *Prodenia exempta* Walker, 1857; *Prodenia ingloria* Walker, 1858; *Laphygma exempta* Hampson, 1909.

S. exempta feeds on finger millet, sorghum, tef, pearl millet, and little millet in Africa and Asia.

Red tef worm, *Mentaxya ignicollis* (Walker, 1857). Original species combination: *Agrotis ignicollis* Walker, 1857: 740; *Agrotis mesosema* Hampson, 1901: 417; *Lycophotia ignetincta* Hampson, 1918: 114. This is a serious pest feeding on tef in Africa.

Subfamily: Heliothinae

Cotton bollworm, *Helicoverpa armigera* (Hübner, 1809). Original species combination: *Chloridea armigera* Hübner, 1809. Syn. *Heliothis armigera* Hübner, 1805; *Chloridea obsoleta* Duncan and Westwood, 1841; *Helicoverpa commoni* Hardwick, 1965. This species feeds on earhead of sorghum and finger millet in India.

Millet head miner moth, *Heliocheilus albipunctella* (de Joannis, 1925). Original species combination: *Raghuva albipunctella* de Joannis, 1925. This is a serious pest on pearl millet in sahelian Africa.

Subfamily: Acontiinae

***Autoba silicula* (Swinhoe, 1897)**

Syn. *Eublemma silicula* Swinhoe, 1897: 167; *Autoba silicula saturata* Warren, 1913: 225; *Eublemma compsoprepes* Turner, 1945: 157. This species forms a web in the earhead and feeds on grains.

Subfamily: Aganainae

***Asota plaginota* (Butler, 1875) (Fig. 2.11)**. Original species combination: *Hypsa plaginota* Butler, 1875. This moth feeds on millets in northeast India.

Subfamily: Catocalinae

***Mocis frugalis* (Fabricius, 1775)**

FIGURE 2.11

Adult moth of *Asota plaginota*.

Courtesy of A. Kalaisekar.

Syn. *Noctua frugalis* Fabricius, 1775; *Chalciope lycopodia* Geyer, 1837; *Remigia translata* Walker, 1865; *Remigia nigripunctata* Warren, 1913

This is a semilooper and it feeds on leaves as well as developing grains in Asia and Africa.

Family: Pyralidae
Subfamily: Phycitinae

Honey dew moth, *Cryptoblabes gnidiella* **(Millière, 1867)**

Syn. *Ephestia gnidiella* Millière, 1867: 308

Larvae feed on the earhead and leaves of sorghum. Adult moths are associated with coccids for honeydews (Ben-Shaul et al., 1991–92).

MIDGES

Order: Diptera; Suborder: Nematocera; Superfamily: Sciaroidea
Family: Cecidomyiidae
Genus: *Stenodiplosis* **Reuter, 1895**
Sorghum midge, *Stenodiplosis sorghicola* **(Coquillett, 1899) (Fig. 2.12)**

Original species combination: *Diplosis sorghicola* Coquillett, 1898: 82. Syn. *Contarinia sorghicola* (Coquillett)

FIGURE 2.12

Adult fly of *Stenodiplosis sorghicola*.

Courtesy of A. Kalaisekar.

This species is a native of Africa (Bowden and Neve, 1953; Young and Teetes, 1977) and spread to other parts of the world. Coquillett first described the species as *D. sorghicola* in 1898. The adult body length is 2 mm. The antenna is composed of 14 joints; the antenna in the male is as long as the body, with bristles arranged in whorls; in the female it is one-half as long as the body, with bristles not arranged in whorls and scattered. The head including the palpi is yellow; antenna and legs are brown; thorax and abdomen are orange red; the center of the mesonotum is black; a black spot crosses the pleura and enlarges on the sternum. The wings are grayish hyaline, the first vein reaches the costa before the middle of the wing; third vein is nearly straight, ending slightly below the extreme tip of the wing; fifth vein is forked slightly before the middle of the wing. The female possesses a delicate telescopic ovipositor/circus. Characters of the female cerci are especially useful in diagnosing species.

Millet midge *Geiromiya penniseti* (Felt, 1920). Original species combination: *Itonida pennisetti* Felt, 1920.

This midge species is reported from Africa and India (Sharma and Davies, 1988) as a pest on developing grains of pearl millet during the rainy season. Detailed biology is established for this species (Coutin and Harris, 1968). But the taxonomic identity needs to be studied further.

APHIDS

Order: Hemiptera; Suborder: Sternorrhyncha; Superfamily: Aphidoidea.
Family: Aphididae
Subfamily: Aphidinae
Sugarcane aphid, *Melanaphis sacchari* (Zehntner, 1897) (Fig. 2.13)
Syn. *Aphis sacchari* (Zehntner, 1897); Zimmerman, 1948; *Longiunguis sacchari* (Zehntner, 1897); Eastop, 1965: 399–592.

FIGURE 2.13

Alate and wingless forms of *Melanaphis sacchari*.

Courtesy of A. Kalaisekar.

This is a key pest on sorghum and sugarcane in many areas of Africa, Asia, Australia, the Far East, and parts of Central and South America (Singh et al., 2004). The identification characters are as follows (Pérez Hidalgo et al., 2012):

- aphids that are broad and oval-shaped
- siphunculus shaped differently and with functional aperture
- siphunculus tubular and shorter than cauda, usually with a well-developed swollen flange and without apical zone of polygonal reticulation
- cauda with 7–20 setae and longest on hind tibia; coxae pale
- dorsum usually without pigmentation.
- head without spicules and front of head without horns
- no evident wax glands
- alatae viviparous females with wing veins dark bordered
- abdominal segments I and VII with marginal papillae dorsal to the respective spiracular apertures
- dorsum of abdomen with membranous cuticle

Corn aphid, *Rhopalosiphum maidis* (Fitch, 1856) (Fig. 2.14)

Syn. *Aphis adjusta* Zehntner, 1897; *Aphis africana* Theobald, 1914; *Aphis cooki* Essig, 1911; *Aphis maidis* Fitch, 1856; *Aphis maydis* del Guercio, 1913; *Aphis vulpiae* del Guercio, 1913; *Aphis zeae* Bonafous, 1835; *Rhopalosiphum zeae* Rusanova, 1960; *Schizaphis setariae* Rusanova, 1962; *Stenaphis monticellii* del Guercio, 1913

The corn aphid is distributed throughout the world and infests maize and sorghum. It transmits maize dwarf mosaic virus. High populations are noticed during booting stage and heavy earhead infestation just prior to harvest creates problems in harvesting because of honeydew secretion (Singh et al., 2004).

The virgin winged female has head, thorax, antenna, legs, and siphunculi that are brownish black; yellow-green abdomen with brown marginal plates; antenna with secondary rhinariums on segments

FIGURE 2.14

Wingless forms of *Rhopalosiphum maidis*.

Courtesy of A. Kalaisekar.

3–5; shorter siphunculus; bifurcated medial vein on forewing, very short second branch, beginning near the wing margin.

The ovoid apterous female has a gray-green or white-green body; head, legs, and siphunculus that are black; seven or eight abdominal tergites with dark transversal stripes; antenna without secondary rhinariums; the beak reaches to midcoxae; cylindrical siphunculus is short with a swelling at the apex and with a constriction; marginal tubercles are small. The eggs are black and oval. In colder regions both eggs and imagoes (apterous parthenogenetic females) overwinter on cereal grasses (Foott, 1977).

Following are the aphid species recorded on millets whose host information is given in Chapter 4.

Anoecia **spp.**

These species are root feeders on many cereals and millets.

    Anoecia corni **(Fabricius, 1775)**. Original species combination: *Aphis corni* Fabricius 1775

    Syn. *Anoecia agrostidis* Börner, 1950; *Schizoneura corni* Hartig, 1841; *Anoecia cornii* Chakrabarti, Ghosh and Chowdhuri, 1970; *Anoecia disculigera* Börner, 1950; *Schizoneura graminis* Del Guercio, 1895; *Schizoneura obscura* Walker, 1852

    Alatae have a characteristic dark posterodorsal abdominal patch and a large black pterostigmal spot on the forewing. Apterae are greenish gray or gray with a dark sclerotic dorsal abdominal plate. Young nymphs are white or cream in color (Blackman and Eastop, 2000).

    Apterae on grass roots are greenish gray to brown with sclerotized parts dark gray. Apterae in spring on cornus are dark brown. Fundatrices have reduced, three-faceted eyes and five-segmented antennae, whereas subsequent generations have large compound eyes and six-segmented antennae.

    Anoecia cornicola **(Walsh, 1863)**. Original species combination: *Eriosoma cornicola* Walsh, 1863, 1:304

    Syn. *Anoecia querci* (Fitch, 1859); *Anoecia eleusinis* (Thomas, 1878); *Anoecia fungicola* (Walsh, 1863); *Anoecia panicola* (Thomas, 1879)

    Recorded on *S. bicolor, S. glauca, S. viridis, Echinochloa crus-galli, Z. mays, Eragrostis major, Digitaria sanguinalis, Panicum capillare* in the United States.

    Anoecia (Anoecia) fulviabdominalis Eastop and Hille Ris Lambers, 1976

    Anoecia krizusi **(Börner, 1950)**. Original species combination: *Neanoecia krizusi* Börner, 1950.

    Anoecia vagans **(Koch, 1856)**. Original species combination: *Schizoneura vagans* Koch, 1856, 8: 268. Syn. *Anoecia (Anoecia) cerealium* Börner, 1952; *Anoecia (Anoecia) kochii* Schouteden. 1906; *S. corni*; *Schizoneura cerealium* Szanilso, 1880; *Anoecia (Anoecia) radicicola* Eastop and Hille Ris Lambers, 1976; *Colopha rossica* Cholodkovsky, 1898; *Schizonevra venusta* Passerini, 1860; *A. corni* variety *viridis* Börner and Blunk, 1916

Aphis fabae **Scopoli, 1763**
Aphis gossypii **Glover, 1877**
Aphis spiraecola **Patch, 1914**

Syn. *Aphis (Aphis) bidentis* Eastop and Hille Ris Lambers 1976, 45: 49; *Aphis (Aphis) citricola* Essig and Kuwana. 1918, 48(3): 68; *Aphis croomiae* Shinji, 1922, 34(407): 794; *Aphis deutziae* Shinji, 1922, 34(407): 795; *Anuraphis erratica* Del Guercio, 1917, 12(1–2): 221, 233; *Aphis eupatorii* Oestlund, 1886, 14: 39; *Aphis malvoides* van der Goot, 1917, 1(3): 90; *Aphis malvoïdes* van der Goot, 1917, 1(3): 90; *Aphis mitsubae* Shinji, 1922, 34(407): 799; *Aphis nigricauda* van der Goot, 1917, 1(3): 91; *Aphis* *nostras* Hottes, 1930, 43: 180; *Aphis pirifoliae* Shinji, 1922, 34(407): 798; *Aphis pseudopomi* Blanchard, 1939, 17: 911, 926; *Aphis viburnicolens* Swain, 1919, 3: 89

    ***Brachycaudus helichrysi* (Kaltenbach, 1843)**. Original species combination: *Aphis helichrysi* Kaltenbach, 1843: 102

Sugarcane woolly aphid, *Ceratovacuna lanigera* Zehntner, 1897

Syn. *Oregma lanigera* van Deventer, 1906; *Cerataphis saccharivora* Matsumura, 1917

    Apterae are light-colored pale green or brownish yellow or grayish brown, densely covered with white waxy filaments along the margins of the body. Alatae are brownish or black. This species is found in India (since 2002) and southeast Asia.

    ***Diuraphis noxia* (Kurdjumov, 1913)**. Original species combination: *Brachycolus noxia* Kurdjumov, 1913

    Apterae are pale yellow or yellowish green or grayish green and lightly dusted with white wax powder. This species is recorded on tef in Africa.

Forda formicaria von Heyden, 1837, 2(3): 292

Syn. *Forda intermixta* Börner, 1952; *Forda meridionalis; Forda occidentalis* Hart, 1894; *Forda (Forda) formicaria subnuda* Börner, 1952; *Rhizoterus vacca* Hartig, 1841; *Forda viridana* Buckton, 1883; *Pentaphis viridescens* Del Guercio, 1921

    Recorded on barnyardgrass, *Elymus* sp., *Hordeum* spp., *Setaria* spp., wheat, and oats in the United States.

Forda marginata Koch, 1857

Syn. *Pentaphis apuliae* Del Guercio, 1921; *Pemphigus follicularias* Macchiati, 1881;

    Forda follicularioides Mordvilko, 1935; *Pemphigus follicularius* Passerini, 1861; *Forda interjecta* Cockerell, 1903; *Forda kingii* Cockerell, 1903; *Forda mokrzeckyi* Mordvilko, 1921; *Forda olivacea* Rohwer, 1908; *Forda polonica* Mordvilko, 1921; *Forda proximalis* Mordvilko, 1921; *Pemphigus retroflexus* Courchet, 1879; *Tychea trivalis* Zhang, Qiao and Chen, 1999; *Tychea trivialis* Passerini, 1860; *Forda wilsoni* Mordvilko, 1921

    Recorded on barnyardgrass, *Elymus* sp., *Hordeum* spp., *Setaria* spp., wheat, and oats in the United States.

Forda hirsuta Mordvilko, 1935

Forda orientalis David, 1969

Syn. *Forda ussuriensis* Mordvilko, 1935

    ***Geoica lucifuga* (Schouteden, 1907)**. Original species combination: *Tetraneura lucifuga* Schouteden, 1907, 50: 34. Syn. *Geoica golbachi* Blanchard, 1958, 15: 155; *Geoica horvathi* Nevsky, 1929, 82: 224; *Geoica pseudosetulosa* Theobald, 1928, 19: 179; *Geoica spatulata* Theobald, 1923, 7: 73

    ***Geoica utricularia* (Passerini, 1860)**. Original species combination: *Pemphigus utricularius* Passerini, 1860

    Mealy plum aphid, *Hyalopterus pruni* (Geoffroy 1762). Original species combination: *Aphis pruni* Geoffroy, 1762

    Alatae with green abdomen and white wax patches segmentally. Apterae are elongated with darker green mottling.

Rusty plum aphid, *Hysteroneura setariae* (Thomas, 1878)

Original species combination: *Siphonophora setariae* Thomas, 1878, 1(2): 5. Syn. *Hysteroneura bituberculata* Börner, 1926; *Hysteroneura gregoryi* Remaudière, G. and M. Remaudière, 1997; *Hysteroneura ogloblini* Blanchard, 1939; *Hysteroneura panicola* Monell, 1879; *Nectarophora prunicola* Hunter, 1901; *Aphis prunicolens* Hunter, 1901; *Hysteroneura prunicoleus* Hunter. 1901; *Hysteroneura scotti* Patch, 1923; *Aphis spiraeae* Oestlund, 1887. This species is native to the United States.

These aphids feed on sorghum and other millets in the United States. Apterae have a rusty brown body. Segments of the antennae and tibiae are contrastingly pale. They have black siphunculi and pale cauda.

Melanaphis pyraria (Passerini, 1861). Original species combination: *Myzus pyrarius* Passerini, 1861. Syn. *Geoktapia 5-articulata* Börner, 1952; *Geoktapia areshensis* Mordvilko, 1921; *Myzus pirarius* Del Guercio, 1900; *Myzus pirinus* Del Guercio, 1900; *Aphis pyrastri* Boisduval, 1867; *Myzus pyrinus* Ferrari, 1872; *Schizaphidiella quinquarticulata* Hille Ris Lambers, 1939; *Melanaphis streili* Börner, 1952

Melanaphis sorghi (Schouteden, 1907). Original species combination: *Aphis sorghi* Schouteden, 1907.

Rose-grass aphid, *Metopolophium dirhodum* (Walker, 1849)

Syn. *Aphis dirhoda* Walker, 1849; *Acyrthosiphon (Metopolophium) dirhodum* (Walker, 1849); *Siphonophora dirhoda* (Walker, 1858); *Siphonophora dirhodum* Buckton, 1876; *Siphonophora longipennis* (Buckton, 1876); *Macrosiphon dirhodum* Schouteden, 1901; *Goidanichiella graminum* (Theobald, 1913); *Macrosiphum arundinis* (Theobald, 1913)

A. dirhoda Walker, 1849, was the type specimen on which the genus *Metopolophium* Mordvilko, 1914, was erected. Taxonomic identification keys are available for this genus (Prior, 1976; Stroyan, 1984). Apterae are green or yellowish green with a brighter green spinal stripe. Each segment of antenna is with dark apices (Blackman and Eastop, 2000). This species is distributed throughout the world and recorded on finger millet.

Myzus persicae (Sulzer, 1776). Original species combination: *Aphis persicae* Sulzer, 1776

Neomyzus circumflexus (Buckton, 1876). Original species combination: *Siphonophora circumflexa* Buckton, 1876. Syn. *Neomyzus callae* Mason, 1940; *Macrosiphum primulanum* Matsumura, 1917; *Myzus vincae* Gillette, 1908

Paracletus cimiciformis von Heyden, 1837.

Protaphis middletonii **(Thomas, 1879)**. Original species combination: *Aphis middletonii* Thomas, C. 1879. Syn. *Aphis armoraciae* Cowen, 1895; *Aphis madiradicis* Knowlton and Smith, 1936; *Protaphis maidiradicis* Forbes, 1894; *Protaphis maidi-radicis* Davis, 1909; *Aphis maydiradicis* Del Guercio, 1914; *Aphis menthaeradicis* Cowen, 1895; *Aphis menthae-radicis* Cowen, 1895; *Aphis middletoni* Hunter, 1901

Rhopalosiphum nymphaeae **(Linnaeus, 1761)**. Original species combination: *Aphis nymphaeae* Linnaeus, 1761. Syn. *Rhopalosiphum acuaticus* Carrillo, 1980; *Rhopalosiphum alismae* Koch, 1854; *Rhopalosiphum aquatica* Theobald, 1915; *Aphis butomi* Schrank, 1801; *Aphis infuscata* Koch, 1854; *Rhopalosiphum naiadum* Walker, 1858; *Rhopalosiphum najadum* Koch, 1854; *Aphis nymphäae* Kaltenbach, 1843; *Rhopalosiphum nymphae* Knowlton, 1935; *Aphis nympheae* Fabricius, 1794; *Rhopalosiphum nympheas* Verma and Das, 1992; *Aphis plantarum aquaticarum* Fabricius, 1794; *Rhopalosiphum plantarumaquaticum* Eastop and Hille Ris Lambers, 1976; *Rhopalosiphum plantarum-aquaticum* Carrillo, 1980; *Aphis prunaria* Walker, 1848; *Rhopalosiphum prunorum* Patch, 1914; *Rhopalosiphum sparganii* Hille Ris Lambers, 1934; *Rhopalosiphum yoksumi* Ghosh, Banerjee and Raychaudhuri, 1971

***Rhopalosiphum padi* (Linnaeus, 1758)**. Syn. *Siphocoryne acericola* Matsumura, 1917; *Siphona-phis padi americana* Mordvilko, 1921; *Aphis avenae sativae* Schrank, 1801; *Aphis avenaesativae* Schrank, 1801; *Rhopalosiphon avenae-sativae* Börner, 1952; *Siphocoryne donarium* Matsumura, 1918; *Siphocoryne fraxinicola* Matsumura, 1917; *Aphis holci* Ferrari, 1872; *Aphis prunifoliae* Fitch, 1855; *Aphis prunifolii* Monell, 1879; *Aphis pseudoavenae* Patch, 1917; *Aphis avenae sativae* Schrank, 1801; *Rhopalosiphum tritici* Hunter, 1901; *Aphis uwamizusakurae* Monzen, 1929

***Rhopalosiphum rufiabdominale* (Matsumura, 1917)**. Original species combination: *Yamata-phis rufiabdominalis* Matsumura, 1917. Syn. *Cerosipha californica* Essig, 1944; *Rhopalosiphum fucanoi* Remaudière, G. and M. Remaudière, 1997; *Rhopalosiphum gnaphalii* Tissot, 1933; *Anura-phis mume* Hori, 1927; *Yamataphis oryzae* Matsumura, 1917; *Yamataphis papaveri* Takahashi, 1921; *Rhopalosiphum pruni* Doncaster, 1956; *Aresha setigera* Blanchard, 1939; *Aresha shel-kovnikovi* Mordvilko, 1921; *Siphocoryne splendens* Theobald, 1915; *Rhopalosiphum subterra-neum* Mason, 1937

Greenbug, *Schizaphis graminum* (Rondani, 1852). Original species combination: *Aphis grami-num* Rondani, 1852, 6: 9–11

This is a pest of great concern, especially in the United States, and has been recognized as a major pest of small grains for over 150 years. Outbreaks of the greenbug, *S. graminum*, on sorghum in the United States are not uncommon. It first entered the United States from Europe around 1882 (Hunter, 1901). It has long been associated with sorghum and it is a well-established pest on sorghum now.

The spring grain aphid is well known as "greenbug," probably since 1882, when a correspondent of the newspaper *Country Gentleman* from Virginia wrote in a news item that reads "Wheat looking well and promising, but there is a little **green bug** on it that may injure it. This same little green fellow is ruining the oats in this neighborhood, and has already destroyed them entirely in many localities" (sic) (Webster and Phillips, 1912).

Greenbug is a most important insect pest of winter wheat and sorghum in the southern Great Plains of the United States (Starks and Burton, 1977; Webster, 1995; Royer, 2015).

Adults and nymphs are light green and oval shaped. Antennae are dark, cornicles are with dark tips, and it has green legs and dark tarsi.

Yellow sorghum aphis, *Sipha flava* (Forbes, 1884). Syn. *Chaitophorus flava* Forbes, 1884. This species was reported in the United States as a pest on sorghum as early as 1876. *S. flava* from Illinois was described by Forbes in 1884.

Alatae have dark dorsal abdominal markings. Apterae are with dusky or gray transverse interseg-mental markings. The overall body is straw colored or slight to bright yellow or green especially in low temperatures. It causes damage to sorghum, sugarcane, and several species of pasture grass (Medina-Gaud et al., 1965; Kindler and Dalrymple, 1999).

Sipha elegans Del Guercio, 1905

Syn. *Sipha (Rungsia) aegilops* Remaudière, G. and M. Remaudière, 1997; *Sipha agropyrella* Hille Ris Lambers, 1939; *Sipha kurdjumovi* Mordvilko, 1921; *Sipha (Rungsia) nemaydis* Remaudière, G. and M. Remaudière, 1997

Sipha maydis Passerini, 1860

Syn. *Sipha avenae* Del Guercio, 1900; *S. maydis* var. *avenae* Del Guercio, 1905; *Sipha (Rungsia) brun-nea* Remaudière, G. and M. Remaudière, 1997; *Sipha (Rungsia) graminis* Börner, 1952; *S. (Rungsia) graminis* Kaltenbach, 1874; *Sipha maydisv* Zhang, Qiao and Chen, 1999

Sitobion africanum **(Hille Ris Lambers, 1954)**

Sitobion avenae **(Fabricius, 1775)**. Original species combination: *Aphis avenae* Fabricius, 1775

Sitobion graminis **Takahashi, 1950**

Syn. *Macrosiphum (Sitobion) gathaka* Eastop, 1955; *Sitobion javanicum* Noordam, 2004

Sitobion indicum **Basu, 1964**

Sitobion leelamaniae **David, 1976**

Syn. *Macrosiphum (Sitobion) chanikiwiti* Eastop, 1959; *Macrosiphum (Sitobion) howlandae* Eastop, 1959

Indian grain aphid, *Sitobion miscanthi* (Takahashi, 1921)

Syn. *Macrosiphum miscanthi* Takahashi, 1921; *Macrosiphum eleusines* Theobald, 1929; *Sitobion eleusines* Theobald, 1929. This species has long tubular siphunculi, five- or six-segmented antennae longer than its body length, and a tapering tongue-shaped cauda (David, 1975). Apterae are green or reddish brown or dark brown; siphunculi are shiny black, cauda are pale, and dorsal cuticle is tanned with intersegmental sclerites.

This species feeds on finger millet, sorghum, and pearl millet in Asia.

Sitobion pauliani **Remaudière, 1957**

Smynthurodes betae **Westwood, 1849**

Syn. *Amycla albicornis* Koch, 1857; *Pemphigus globosus* Walker, 1852; *Smynthurodes gossypii* Channa Basavanna, 1962, 24: 85; *Smynthurodes karschii* Börner, 1952; *Aphis myrmecaria* Boisduval, 1867; *Forda natalensis* Theobald, 1920; *Trifidaphis perniciosus* Nevsky, 1929; *Tychea phaseoli* Passerini, 1860; *Trifidaphis racidicola* Wilson, 1910; *Pemphigus radicicola* Essig, 1909; *Geoica radicola* Cutright, 1925; *Trifidaphis silvestrii* Mordvilko, 1935; *Pemphigus trifolii* Del Guercio, 1915

Tetraneura **spp.**

The genus *Tetraneura* has 30 species. The apterae of this genus are usually very globose and have one-segmented tarsi, and the alatae have a simple, unbranched media in the forewing. These species are root feeders.

Tetraneura nigriabdominalis (Sasaki, 1899). Original species combination: *Schizoneura nigriabdominalis* Sasaki, 1899. Syn. *Tetraneura graminiradicis* (Zhang, 1992); *Tetraneura hirsuta* (Baker, 1921). Subgenus and subspecies are recognized for this species (genus *Tetraneura* Hartig, 1841; subgenus *Tetraneurella* Hille Ris Lambers, 1970; subsp. *nigriabdominalis*), therefore the valid name is *Tetraneura (Tetraneurella) nigriabdominalis nigriabdominalis* (Sasaki, 1899).

T. nigriabdominalis is widely distributed in Africa, India, Pakistan, Sri Lanka, Japan, China, Korea, Malaysia, and the Philippines (Blackman and Eastop, 1984).

Tetraneura africana **van der Goot, 1912**

Syn. *Tetraneura cynodontis* Theobald, 1923

Tetraneura caerulescens **Passerini, 1860**

Syn. *Tetraneura aegyptiaca* Theobald, 1923; *Pemphigus coerulescens* Macchiati, 1882; *Tetraneura (Tetraneura) rubra* de Horváth, 1894

Tetraneura chui **Zhang and Zhang, 1991**

Tetraneura ulmi **(Linnaeus, 1758)**. Original species combination: *Aphis ulmi* Linnaeus, 1746. Syn. *Endeis bella* Koch, 1857; *Tetraneura (Tetraneura) boyeri* Passerini, 1861; *Amycla fuscifrons* Koch, 1857; *Aphis gallarum* Gmelin, 1790; *Aphis gallarum ulmi* De Geer, 1773; *Aphis gallarumulmi* De Geer, 1773; *Aphis gallarum-ulmi* Goeze, 1778; *Pemphigus graminis* Schouteden, 1906; *Byrsocrypta personata* Börner, 1950; *Aphis radicum* Boyer de Fonscolombe, 1841; *Endeis rorea* Koch, 1857; *Endeis rosea* Koch, 1857; *Pemphigus fuscifrons* var. *saccarata* Del Guercio, 1895; *Tetraneura saccharata* Schouteden, 1907; *Tetraneura (Tetraneura) theobaldi* Remaudière, G. and M. Remaudière 1997; *Tetraneura ulmifoliae* Baker, 1920; *Tetraneura (Tetraneura) ulmigallarum* Remaudière, G. and M. Remaudière, 1997; *Tetraneura ulmisacculi* Patch, 1910; *Pemphigus zeae maydis* Cholodkovsky, 1902; *Tetraneura (Tetraneura) zeae-maidis* Börner, 1952; *Tetraneura (Tetraneura) zeaemaydis* Walker, 1852; *Pemphigus zeamaidis* Macchiati, 1883

Tetraneura basui **Hille Ris Lambers, 1970**

Tetraneura yezoensis **Mordvilko, 1921**

Syn. *Tetraneura heterohirsuta* Carver and Basu, 1961

Tetraneura javensis **van der Goot, 1917**

Syn. *Tetraneura (Indotetraneura) coimbatorensis* Hille Ris Lambers, 1970

Tetraneura capitata **Zhang and Zhang, 1991.**

Tetraneura triangula **Zhang and Zhang, 1991.**

BUGS

Order: Hemiptera; Suborder: Auchenorrhyncha; Infraorder: Cicadomorpha; Superfamily: Membracoidea

Family: Cicadellidae

Maize leafhopper, *Cicadulina mbila* **(Naudé, 1924)**. Original combination: *Balclutha mbila* Naudé, 1924. *C. mbila* (Naudé, 1924) China, 1928: 61–63.

The reliable diagnostic characters for species identification in the genus *Cicadulina* are aedeagus and pygophore processes of the male genitalia (Webb, 1987). This species is reported from India on pearl millet, sorghum, finger millet, and other small millets. It is distributed in India, Central Asia, and Africa. Other important species of this genus are *Cicadulina storey, Cicadulina bipunctella bipunctella,* and *Cicadulina chinai.*

Rice green leafhopper, *Nephotettix cincticeps* **(Uhler, 1896)**. Original species combination: *Selenocephalus cincticeps* Uhler, 1896. Syn. *Nephotettix apicalis* Melichar; *N. apicalis cincticeps* Esaki and Hashimoto; *Nephotettix bipunctatus cincticeps* Esaki and Hashimoto

Taxonomic revisions on this genus clarified the identity of many species (Ghauri, 1971). *N. cincticeps* has been recorded on barnyard millet. Other important species of this genus are *Nephotettix nigropictus* and *Nephotettix virescens.*

Leafhopper, *Empoasca flavescens* **(Fabricius, 1794).** Original species combination: *Cicada flavescens* Fabricius, 1794: 46. This species is found in the Asian region on sorghum and other millets. **Infraorder: Fulgoromorpha; Superfamily: Fulgoroidea**

Family: Delphacidae

Shoot bug, *Peregrinus maidis* **(Ashmead, 1890)**

(Staphylininae: *Peregrinus* **Kirkaldy, 1904)**

Original species combination: *Delphax maidis* Ashmead, 1890: 167–168. **Syn.** *Delphax psylloides* Lethierry, 1897: 105–106; *Dicranotropis maidis* (Ashmead, 1890); van Duzee 1897: 225–261; *Liburnia psylloides* (Lethierry, 1890); Kirkaldy 1904: 175–179; *Peregrinus maidis* (Ashmead, 1890); Kirkaldy 1907: 186.

This species causes economic losses in maize and sorghum (Dupo and Barrion et al., 2009) in the tropics and coastal areas of subtropical and temperate regions of all continents (Singh and Seetharama, 2008). *P. maidis*-vectored maize mosaic virus (MMV) wiped out the maize crop of the Maya civilization in Central America and thus was considered to be one of the main reasons for the classic Maya collapse (Brewbaker, 1979; Hannikainen, 2011). But there are rebuttals to the claims of *P. maidis* being the prime architect of the collapse, because *P. maidis* is the sole vector of MMV and maize stripe virus and *Sorghum* spp. are the ancestral plant hosts. The vector and viruses were adapted to maize in post-Columbian times and this is in conflict with the hypothesis that *P. maidis* and MMV mediated the classic Maya collapse in pre-Columbian times (Nault, 1983). Nevertheless, there ought to be claims and counterclaims in the course of any historical evidence-based scientific hypothesis. The irrefutable fact is that *P. maidis* has been a pest of serious concern to sorghum and maize for centuries in many countries.

Innumerous nymphs and adults feed within the leaf sheath (Fig. 2.15). The adult bugs show macropterous and brachypterous winged forms. In the macropterous wing, the membranous region is marked with radiating bands and a dark patch at the middle of the anal margin. Brachypterous wings show a distal thick dark patch and a small anal spot.

FIGURE 2.15

Nymphs and adults of *Peregrinus maidis*.

Courtesy of A. Kalaisekar.

Brown plant hopper, *Nilaparvata lugens* **(Stål, 1854)**. Original combination: *Delphax lugens* Stål, 1854: 246. Syn. *Delphax oryzae* Matsumura, 1906: 13; *Nilaparvata greeni* Distant, 1906: 486. This plant hopper feeds on millets in Asia.

Plant hopper, *Sogatella furcifera* **(Horváth, 1899)**. Original species combination: *Delphax furcifera* Horváth, 1899: 372. This plant hopper feeds on sorghum in Asia and the Middle East (Asche and Wilson, 1990).

Family: Fulgoridae

Plant hopper, *Proutista moesta* **(Westwood, 1896)**. Original species combination: *Derbe (Phenice) moesta* Westwood, 1896.

Family: Lophopidae

Sugarcane leafhopper, *Pyrilla perpusilla* **(Walker, 1851).** Original species combination: *Pyrops perpusilla* Walker, 1851: 269. It feeds on sorghum in Asia.

Suborder: Sternorrhyncha; Superfamily: Coccoidea

Family: Pseudococcidae

Rice mealybug, *Brevennia rehi* **(Lindinger)**

B. rehi (Lindinger, 1943) Miller, 1973: 372.

Syn. *Ripersia sacchari* Green, 1931; *Ripersia rehi* Lindinger, 1943; *Tychea rehi* Lindinger, 1943; *Heterococcus rehi* (Lindinger, 1943) Williams,1970; *Heterococcus tuttlei* Miller and McKenzie, 1970. This species is recorded on sorghum in India.

Suborder: Heteroptera; Infraorder: Pentatomorpha

Superfamily: Pentatomoidea

Family: Cydnidae

Burrowing bugs *Stibaropus* Dallas, 1851. The following species of this genus suck the sap from roots underground: *Stibaropus callidus* **(Schiodte), 1848,** and *Stibaropus molginus* **(Schiodte), 1848.** They feed on sorghum in India.

Superfamily: Coreoidea

Family: Alydidae

Paddy earhead bug, *Leptocorisa acuta* **(Thunberg, 1783).**

Original species combination: *Cimex acutus* Thunberg, 1783: 34. Syn. *Leptocorisa angustatus* (Fabricius, 1787); *Leptocorisa arcuata* (Kolenati, 1845); *Leptocorisa varicornis* (Fabricius, 1803). This species feeds on sorghum, pearl millet, and finger millet in Asia.

Riptortus linearis **(Fabricius, 1775).** Original species combination: *Cimex linearis* Fabricius, 1775. This species feeds on sorghum in India.

Family: Coreidae

Rice stink bug, *Cletus punctiger* **(Dallas, 1852)** (Fig. 2.16). Original species combination: *Gonocerus punctiger* Dallas, 1852: 494

The species of the genus *Cletus* have an oblong and subcompressed body and acutely produced lateral spines of the pronotum (Rashmi and Devinder, 2013). These species can be commonly sighted on sorghum and other grass species in India.

Leptoglossus phyllopus **(Linnaeus, 1767)**. Original species combination: *Cimex phyllopus* Linnaeus, 1767: 731. Syn. *Anisoscelis albicinctus* Say, 1832; *Anisoscelis confusa* Dallas, 1852;

Anisoscelis fraterna Westwood, 1842. These insects suck the sap from leaves and developing grains of sorghum, pearl millet, and barnyard millet. There are many other species of *Leptoglossus* recorded on millets (Fig. 2.17). They are distributed worldwide.

Superfamily: Lygaeoidea
Family: Blissidae

FIGURE 2.16

Adult of *Cletus punctiger*.

Courtesy of A. Kalaisekar.

FIGURE 2.17

Adult of *Leptoglossus* sp.

Courtesy of A. Kalaisekar.

Chinch bugs, *Blissus leucopterus* (Say, 1831). Original species combination: *Lygaeus leucopterus* Say, 1831: 329

Chinch (= pest, in Spanish) bugs were reported to be damaging wheat crops as early as 1783 in North Carolina, United States, and during those periods, the bug was a major threat to the agrarian economy, with populations reaching outbreak levels (Leonard, 1966). Owing to the sheer number of the insects and the destructive damage to the wheat, the occurrence of chinch bugs often made headlines of the daily and weekly newspapers for decades in the United States (Leonard, 1966). Chinch bugs were formally described by Say in 1831. They are grass-feeding bugs and are reported on sorghum and millets in the United States.

The bugs are either macropterous or brachypterous in a location. There are some closely similar species and subspecies recognized in the United States, viz., *B. leucopterus leucopterus* (Say), *B. leucopterus hirtus* Montandon, *Blissus insularis* Barber, and *Blissus occiduus* Barber. Although closely related, these chinch bugs have been shown to elicit differential feeding responses in their grass hosts (Anderson et al., 2006).

Family: Lygaeidae

False chinch bug, *Nysius niger* Baker, 1906. This bug feeds on millets of the New World.

Spilostethus pandurus **(Scopoli, 1763)** (Fig. 2.18). Original species combination: *Cimex pandurus* Scopoli, 1763: 126–127. Syn. *Cimex militaris* Fabricius, 1775: 717; *Spilostethus pandurus* Oshamin, 1912, 1: 27; *Spilostethus pandurus militaris* (Fabricius) Mukhopadhyay, 1988, 107: 15.

This species is found in Australia, China, India, the Malayan Archipelago, Myanmar, New Caledonia, Pakistan, Sri Lanka, and the Palearctic region (Ghosh, 2008).

It has a spot at the inner margins of the eyes; a pronotum with two large obscure ochraceous spots on the basal area; a scutellum with a small spot near the apex of the clavus, a transverse fascia to the corium, prosternum, abdominal segmental margins, and with stigmal spots; membrane is black, with a central white spot and two crescent white markings on the membranal suture; labium reaches or passes second coxae; legs are black, femora are distinctly spined beneath (Ghosh, 2008).

This species is found feeding on developing grains of pearl millet in India.

FIGURE 2.18

Adult of *Spilostethus pandurus*.

Courtesy of A. Kalaisekar.

FIGURE 2.19

Adult of *Spilostethus hospes.*

Courtesy of A. Kalaisekar.

Spilostethus hospes **(Fabricius, 1794)** (Fig. 2.19). Original species combination: *Lygaeus hospes* Fabricius, 1794. Syn. *Lygaeus hospes* Fabricius: Distant, 1904, 2: 6; *S. hospes* Bergroth, 1914, 2: 356.

It is found in Australia, China, India, the Malayan Archipelago, Myanmar, New Caledonia, Pakistan, and Sri Lanka (Ghosh, 2008).

It has a spot at the inner margins of the eyes, with two broad central discal fasciae to the pronotum; the rostrum reaches the posterior coxae; scutellum has a large central spot to the corium; membrane, disks of sternum, and abdomen are black; legs are also black, femora are unarmed (Ghosh, 2008).

This species is found feeding on developing grains of pearl millet in India.

Family: Meenoplidae

(Superfamily: Falgoroidea)

Plant hopper, *Nisia atrovenosa* **(Lethierry, 1888).** Original species combination: *Meenoplus atrovenosus* Lethierry, 1888: 460–470. It feeds on sorghum and other millets.

This species has granulated veins in the forewing clavus; at rest, the forewings are folded tentlike over the body. The mirid bug, *Cyrtorhinus lividipennis*, is predaceous on *Nisia* spp.

Family: Miridae

Sorghum head bug, *Calocoris angustatus* **Lethierry, 1893**

(Heteroptera: Cimicomorpha: Miroidea: Miridae: *Calocoris* **Fieber, 1858)**

C. angustatus **Lethierry, 1893, 2: 90.**

C. angustatus was first described by Lethierry in 1893 from the former state of Madras, India (Distant, 1902). The head bug, *C. angustatus* Lethiery, is a key limiting factor in sorghum production (Young and Teetes, 1977; Sharma and Lopez, 1990). Apart from India, the bug has been reported from Kenya and Rwanda (Seshu Reddy and Omolo, 1985).

The adult bug is dull flavescent. It has fulvous antennae, first joint robust, as long as the head; second joint four times as long as the first and slender; third, fourth, and fifth joints equal. The pronotum has trapeziform, punctulate, posterior angles slightly obtusely acute; anterior angles are obtuse, anteriorly with a distinct collar, apices robustly callose; hemelytra is flavescent, punctulate, sparingly fulvous-pubescent; clavus and sutual portion of the corium are roseate; legs are concolorous, tibiae are armed externally with 10 or 11 black spinules; the apices of the tarsi are fuscous.

Other related species of *Calocoris* are ***Calocoris lineolatus* Goeze, 1778**: 267; ***Calocoris dohertyi* Distant, 1904**: 452; and ***Calocoris stoliczkanus* Distant, 1879**: 124.

Geocoris tricolor is predaceous bug on *C. angustatus*.

***Eurystylus bellevoyei* (Reuter, 1879)**. Original species combination: *Eurycyrtus bellevoyei* Reuter, 1879

These bugs are found primarily on plants of Chenopodiaceae. The distribution extends from the Mediterranean to Turkestan, with a wide range in tropical Africa and the Indian subcontinent (Odhiambo, 1958; Carapezza, 1998). It is a serious pest on sorghum earhead in Africa and it is recorded on sorghum in India also (Sharma and Lopez, 1990; Kruger et al., 2007).

***Creontiades pallidus* (Rambur, 1839)**. Original combination: *Phytocoris pallidus* Rambur, 1839

This mirid bug is widely distributed throughout Sudan, Egypt, and Congo (Goodman, 1953) and India (Sharma and Lopez, 1990). It is a serious pest on cotton and recorded on sorghum earheads.

Family: Pentatomidae

Stink bug, *Dolycoris indicus* Stål, 1876. This species feeds on millets in India.

Green stink bug, *Nezara viridula* (Linnaeus, 1758) (Fig. 2.20). Original combination: *Cimex viridula* Linnaeus, 1758. Syn. *Nezara approximata* Reiche and Fairmaire, 1848; *Nezara aurantiaca* Costa, 1884. It feeds on millets in India.

FIGURE 2.20

Ault of *Nezara viridula*.

Courtesy of A. Kalaisekar.

Family: Pyrrhocoreidae

Red cotton bug, *Dysdercus koenigii* Fabricius, 1775. Syn. *Dysdercus cingulatus* Fabricius, 1775: 719. This species lacks ocelli. It is an Asian species that feeds on sorghum and other millets.

Suborder: Sternorrhyncha; Superfamily: Coccoidea. Family: Pseudococcidae

Rice mealybug, *Brevennia rehi* (Lindinger). *B. rehi* (Lindinger, 1943) Miller, 1973: 372. Syn. *Ripersia sacchari* Green, 1931; *Ripersia rehi* Lindinger, 1943; *Tychea rehi* Lindinger, 1943; *Heterococcus rehi* (Lindinger, 1943) Williams,1970; *Heterococcus tuttlei* Miller and McKenzie, 1970. This species is recorded on sorghum and kodo millet in India.

THRIPS

Thripidae: Thysanoptera

The following thrips species are recorded on various millets. Most of the species feed on seedlings.

Chaetanaphothrips orchidii **(Moulton, 1907)**. Syn. *Euthrips orchidii* Moulton, 1907: 52; *Euthrips marginemtorquens* Karny, 1914: 362. Males are not known in this species. It feeds on finger millet and sorghum in India.

Florithrips traegardhi **(Trybom, 1911)**. Syn. *Physapus traegardhi* Trybom, 1911: 63; *Anaphothrips ramakrishnai* Karny, 1926: 203; *Taeniothrips fulvus* Ananthakrishnan and Jagadish, 1969: 114. These thrips feed on millets in India.

Haplothrips aculeatus **F., 1803**. Syn. *Thrips aculeatus* Fabricius, 1803: 312; *Haplothrips cephalotes* Bagnall, 1913: 265. It feeds on sorghum, finger millet, pearl millet, and foxtail millet in Asia. Another important species is *Haplothrips ganglbaueri* **Schmutz, 1913**.

Heliothrips indicus **Bagnall, 1913: 291**. This species feeds on finger millet, sorghum, and pearl millet in India.

Selenothrips rubrocinctus **(Giard, 1901)**. Syn. *Physopus rubrocinctus* Giard, 1901: 263; *Heliothrips (Selenothrips) decolor* Karny, 1911: 179; *Brachyurothrips indicus* Bagnall, 1926: 98. It feeds on millets.

Sorghothrips jonnaphilus **(Ramakrishna, 1928)**. Syn. *Taeniothrips jonnaphilus* Ramakrishna, 1928: 256. It feeds on sorghum in India. Other species of the genus are *Sorghothrips longistylus* (Trybom, 1911), *Sorghothrips fuscus* (Ananthakrishnan, 1965), and *Sorghothrips meishanensis* Chen, 1977.

Stenchaetothrips biformis **(Bagnall, 1913)**. Syn. *Bagnallia biformis* Bagnall, 1913: 237; *Bagnallia melanurus* Bagnall, 1913: 238; *Thrips (Bagnallia) oryzae* Williams, 1916: 353; *Thrips holorphnus* Karny, 1925: 15. These rice thrips feed on sorghum in Asia.

Thrips hawaiiensis **Morgan, 1913**. It feeds on sorghum in Asia.

BEETLES AND WEEVILS

Coleoptera

Chrysomelidae
Flea beetles
Alticinae

Altica cyanea **(Weber, 1801)**. Original species combination: *Haltica cyanea* Weber, 1801. Syn. *Altica birmanensis* Jacoby, 1896. It feeds on sorghum and other millets in India.

   Altica caerulea **(Olivier, 1791)**. Original species combination: *Haltica coerulea* Olivier, 1791. This species feeds on sorghum and other millets in India.

   Chaetocnema basalis **Baly 1877: 310**. Syn. *Chaetocnema geniculata* Jacoby 1896: 270. It is found in Asia and the Palearctic. The genus is thoroughly reviewed for the Palearctic region (Konstantinov et al., 2011). It feeds on sorghum and millets. There are many other species of this genus reported as pests.

   Phyllotreta **spp.** (genus: *Phyllotreta* Chevrolat, 1836: 391). Many species of this genus are recorded on sorghum and millets.

Leaf beetles

Crioccerinae

Cereal leaf beetle, *Oulema melanopus* (Linnaeus, 1758). Original species combination: *Chrysomela melanopus* Linnaeus, 1758: 376. It feeds on sorghum and millets.

Galerucinae

Aulacophora foveicollis **(Lucas, 1849)**. Original species combination: *Galeruca foveicollis* Lucas, 1849. It feeds on millets in India. Another species is *Aulacophora signata* Kirsch, 1866.

Hispinae

Rice hispa, *Dicladispa armigera* (Olivier, 1808). It feeds on sorghum in India.

Coccinellidae

Tef epilachna beetle, *Chnootriba similis* (Thunberg, 1781)

Original species combination: *Coccinella similis* Thunberg, 1781: 15. It is distributed in Africa (Andrzej and Piotr, 2003) and defoliates tef leaves.

Meloidae

Blister beetles

Cylindrothorax tenuicollis (Pallas, 1798). Original species combination: *Meloe tenuicollis* Pallas, 1798: 97–104. Syn. *Cantharis tenuicollis* (Pallas, 1798); *Epicauta tenuicollis* (Pallas,1798); *Lytta tenuicollis* (Pallas,1798). It feeds on sorghum in Asia.

   Mylabris pustulata **(Thunberg, 1791)** (Fig. 2.21). Original species combination: *Meloe pustulata* Thunberg, 1791: 113. It feeds on sorghum in India.

Scarabaeidae

White grubs, *Phyllophaga* spp. and *Holotrichia* spp.

There has been intense disagreement and confusion over the use of generic names such as *Lachnosterna, Phyllophaga, Holotrichia, Melolantha, Ancylonycha,* and *Rhizotrogus* since 1837. The available information on these generic studies largely point to a two-genera theory, which is the use of *Phyllophaga* and *Holotrichia* for the American and the Asian species, respectively. But, further seemingly consolidated ideas are emerging from taxonomists toward adapting a single generic name, *Rhizotrogus*.

   Some of the species of these genera are recorded on finger millet and pearl millet in Africa and Asia.

Cetoniinae

Chauffer beetles

FIGURE 2.21

Adult beetle of *Mylabris pustulata*.

Courtesy of A. Kalaisekar.

***Chiloloba acuta* Wiedemann, 1823** (Fig. 2.22). It is a flower feeder on pearl millet and sorghum in India.

Oxycetonia versicolor (***Fabricius, 1775***). It is a flower feeder on millets in India.

***Torynorrhina flammea* (Gestro, 1888)** (Fig. 2.23). Original species combination: *Rhomborrhina flammea* Gestro, 1888. There are some subspecies designated for this taxon. It feeds on sorghum in India.

Tenibionidae
False wireworms

***Gonocephalum* Solier, 1834**. The following species of this genus are reported on sorghum and other millets in India. These species are present in Africa, but are not reported as pests of millets yet (Ferrer, 2010).

Gonocephalum hoffmannseggii (Steven, 1829). Original species combination: *Opatrum hoffmannseggii* Steven, 1829. Syn. *Dasus hoffmannseggii* (Steven, 1829). This species is widespread in India. Other important species are *Gonocephalum dorsogranosum* (Fairmaire, 1896), *Gonocephalum consobrinum* Blair, 1923, *Gonocephalum tuberculatum* (Hope, 1831), and *Gonocephalum depressum* Fabricius, 1798.

Curculionidae
Ash weevil

Genus: *Myllocerus* Schoenherr, 1823; subgenus: *Myllocerus* Schoenherr, 1823; species: *undecimpustulatus* Faust, 1891: 266; subspecies: *maculosus* Desbrochers, 1899: 111.

FIGURE 2.22

Adult of *Chiloloba acuta*.

Courtesy of A. Kalaisekar.

FIGURE 2.23

Adult of *Torynorrhina flammea*.

Courtesy of A. Kalaisekar.

M. undecimpustulatus maculosus feeds on leaves of sorghum, pearl millet, and finger millet in India.

Argentine stem weevil, *Listronotus bonariensis* (Kuschel, 1955). Original species combination: *Hyperodes bonariensis* Kuschel, 1955. Syn. *Neobagous setosus* Hustache, 1929; *Hyperodes griseus* Marshall, 1937. This weevil species is reported on finger millet and distributed in South America, New Zealand, and Australia.

Leaf weevil, *Tanymecus indicus* Faust, 1895. It feeds on pearl millet and sorghum in India.

GRASSHOPPERS

Acrididae

Acrida exaltata **(Walker, 1859)**. Original species combination: *Truxalis exaltata* Walker, 1859: 222. It feeds on sorghum in Asia.

Aiolopus longicornis **Sjöstedt, 1910: 169**. This species feeds on sorghum and other millets in Africa. Other important species are *Aiolopus simulatrix* and *Aiolopus thalassinus*.

Diabolocatantops axillaris **(Thunberg, 1815)**. Original species combination: *Gryllus axillaris* Thunberg, 1815: 250. It feeds on millets.

Hieroglyphus banian **(Fabricius, 1798)**. Original species combination: *Gryllus banian* Fabricius, 1798: 195. It feeds on sorghum, pearl millet, and finger millet in India.

Hieroglyphus nigrorepleptus **Bolívar, 1912** (Fig. 2.24). This species is the most devastating pest of sorghum and pearl millet in western India.

Hieroglyphus daganensis **Krauss, 1877**. It feeds on sorghum in India.

Locusta migratoria **(Linnaeus, 1758)**. Original species combination: *Gryllus migratorius* Linnaeus, 1758: 432. It feeds on millets in India.

FIGURE 2.24

Adult of *Hieroglyphus nigrorepleptus*.

Courtesy of A. Kalaisekar.

Red locust, *Nomadacris septemfasciata* (Audinet-Serville, 1938). Original species combination: *Acridium septemfasciatum* Audinet-Serville, 1938. Syn. *Acridium coangustatum* Lucas, 1862; *Cyrtacanthacris fascifera* Walker, 1870; *Cyrtacanthacris purpurifera* Walker, 1870; *Cyrtacanthacris subsellata* Walker, 1870; *Cyrtacanthacris septemfasciata* Kirby, 1902; *Acridium fasciferum* Finot, 1907; *Acridium purpuriferum* Finot, 1907; *Acridium sanctae-mariae* Finot, 1907; *Acridium subsellatum* Finot, 1907; *Cyrtacanthacris coangustata* Kirby, 1910; *Cyrtacanthacris sanctae-mariae* Kirby, 1910; *Nomadacris septemfasciata insularis* Chopard, 1936; *Nomadacris fascifera* Orian, 1957; *Patanga septemfasciata* Jago, 1981. The species feeds on sorghum and finger millet in Africa.

Oedaleus senegalensis **(Krauss, 1877).** Original species combination: *Pachytylus senegalensis* Krauss, 1877: 56. It is found in Africa.

Oxya nitidula **(Walker, 1870).** Original species combination: *Acridium nitidulum* Walker, 1870: 631. It feeds on sorghum and pearl millet in India.

Paracinema tricolor **(Thunberg, 1815).** Original species combination: *Gryllus tricolor* Thunberg, 1815: 254.

Schistocerca gregaria **(Forskål, 1775).** Original species combination: *Gryllus gregarius* Forskål, 1775. It is distributed in India and Africa.

Pyrgomorphidae

Atractomorpha crenulata **(Fabricius, 1793).** Original species combination: *Truxalis crenulatus* Fabricius, 1793: 28. Syn. *Atractomorpha obscura* Bolívar, 1917: 392. It feeds on millets in India.

Chrotogonus hemipterus **Schaum, 1853.** It feeds on millets in India.

Tettigoniidae

Conocephalus maculates **(Le Guillou, 1841).** Original species combination: *Xiphidion maculates* Le Guillou, 1841. It has the ability to shoot up its population in a short period and there was a huge swarm of this grasshopper species in the Indian states of Manipur and Nagaland during June–July of 2008. Small millets and maize crops were severely damaged by this species.

Wello-bush cricket, *Decticoides brevipennis* Ragge, 1977. It feeds on tef in Africa.

Armored bush cricket, *Acanthoplus longipes* (Charpentier, 1845). Original species combination: *Hetrodes longipes* Charpentier, 1845. Syn. *Acanthoplus loandae* Péringuey, 1916; *Acanthoplus stratiotes* Brancsik, 1896. It feeds on tef in Africa.

(Taxonomic nomenclatures of Orthopteran species are verified with Eades et al. (2015)).

TERMITES, WASPS AND ANTS

Termites

Termitidae: Isoptera

Odontotermes obesus **(Rambur, 1842),** *Odontotermes microdentatus* **Roonwal and Sen-Sarma, 1960.** A queen termite that was excavated from underground in a field where millets were cultivated in the Khasi hills, Meghalaya, India is shown in Fig. 2.25.

Microtermes obesi **Holmgren, 1913 and** *M. unicolor* **Snyder, 1949.** It feeds on sorghum and pearl millet.

FIGURE 2.25

Queen termite of *Odontotermes* sp.

Courtesy of A. Kalaisekar.

Wasp

Eurytomidae: Hymenoptera

Stem boring wasp, *Eurytomocharis eragrostidis* **Howard, 1896**. Syn. *Eurytoma eragrostidis* (Howard, 1896). This wasp has been recorded infesting *Eragrostis tef* in the United States.

Ants

Family: Formicidae

Subfamily: Myrmicinae

Tribe: Solenopsidini; *Monomorium salomonis* **(Linnaeus, 1758); Tribe: Atini;** *Pheidole sulcaticeps* **Roger, 1863**

REFERENCES

Anderson, W.G., Heng-Moss, T.M., Baxendale, F.P., 2006. Evaluation of cool- and warm-season grasses for resistance to multiple chinch bug (Hemiptera: Blissidae) species. Journal of Economic Entomology 99, 203–211.

Andrzej, S.J., Piotr, W., 2003. World Cataloque of Coccinellidae. Part I. Epilachninae. MANTIS, Olsztyn, Poland. ISBN: 83-918125-3-7.

Asche, M., Wilson, M.R., 1990. The delphacid genus *Sogatella* and related groups: a revision with special reference to rice-associated species (Hornoptera: Fulgoroidea). Systematic Entomology 15, 1–42.

Barrion, A.T., Litsinger, J.A., 1994. Taxonomy of rice insect pests and their arthropod parasites and predators. In: Heinrichs, E.A. (Ed.), Biology and Management of Rice Insects. Wiley Eastern, New Delhi, pp. 13–15. 283–359.

Ben-Shaul, Y., Wysoki, M., Rosen, D., 1991–1992. Phenology of the honeydew moth, *Cryptoblabes gnidiella* (Milliere) (Lepidoptera: Pyralidae), on avocado in Israel. Israel Journal of Entomology 25–26, 149–160.

Blackman, R.L., Eastop, V.F., 1984. Aphids on the World's Crops: An Identification and Information Guide. John Wiley and Sons, Brisbane.

Blackman, R.L., Eastop, V.F., 2000. Aphids on the World's Crops. An Identification and Information Guide, Second ed. John Wiley & Sons, Chichester, p. 414.

Bleszynski, S., 1970. A revision of the world species of *Chilo* Zincken (Lepidoptera: Pyralidae). Bulletin of the British Museum of Natural History (Entomology) 25 (4), 99–195.

Börner, C., 1925. Lepidoptera, Schmetterlinge. pp. 382–421. In: Bröhmer, P. (Ed.), Fauna von Deutschland. Quelle and Meyer, Berlin: Leipzig. 561 p.

Bowden, J., Neve, R.A., 1953. Sorghum midge and resistant varieties in the Gold Coast. Nature 172, 551.

Bradley, J.D., 1981. *Marasmia patnalis* sp. n. (Lepidoptera: Pyralidae) on rice in S.E. Asia. Bulletin of Entomological Research 71 (2), 323–327.

Brewbaker, J., 1979. Diseases of maize in the wet lowland tropics and the collapse of the classic Maya civilization. Economic Botany 33, 101–118.

Carapezza, A., 1998. New species and new records of Heteroptera from Cyprus (Insecta). Attidell'Accademia Roveretana degli Agiati (VII) 8B, 29–40.

Coutin, R., Harris, K.M., 1968. The taxonomy, distribution, biology and economic importance of the millet grain midge, *Geromyia penniseti* (Felt), gen. n., comb. n. (Diptera: Cecidomyiidae). Bulletin of Entomological Research 59 (2), 259–279.

David, S.K., 1975. A taxonomic review of *Macrosiphum* (Homoptera: Aphididae) in India. Oriental Insects 9 (4), 461–493.

Distant, W.L., 1902. The fauna of British India including Ceylon and Burma. In: Rhynchota, vol. 2. Taylor and Francis, London, 503 pp.

Dupo, A.L.B., Barrion, A.T., 2009. Taxonomy and general biology of delphacid planthoppers in rice agro-ecosytems. pp. 3–156. In: Heong, K.L., Hardy, B. (Eds.), Planthoppers: New Threats to the Sustainability of Intensive Rice Production Systems in Asia. International Rice Research Institute, Los Baños (Philippines).

Eades, D.C., Otte, D., Cigliano, M.M., Braun, H., September 30, 2015. Orthoptera Species File. Version 5.0/5.0. http://Orthoptera.SpeciesFile.org.

van Emden, F.I., 1940. Muscidae: B - Coenosiinae. Ruwenzori Expedition 1934–35, 2. British Museum (Natural History), London, UK, pp. 91–255.

Evans, W.H., 1949. A Catalogue of the Hesperiidae from Europe, Asia and Australia in the British Museum (Natural History). Trustees of the British Museum, London.

Fan, Z., 1965. Key to the Common Synanthropic Flies of China. Academy of Sciences, Beijing, China.

Ferrer, J., 2010. Taxonomic notes on the genus *Gonocephalum* Solier, 1834, with description of new taxa (Coleoptera, Tenebrionidae). Annales Zoologici 62, 231–238.

Fletcher, T.B., 1914. Some South Indian Insects and Other Animals of Importance. 565 pp.

Fletcher, T.B., 1917. In: Report of the Proceedings of the Second Entomological Meeting Pusa, India, pp. 47, 53–55, 68, 83, 87, 91, 99, 179, 187, 189, 199, 203.

Fletcher, T.B., 1956. Some South Indian Insects and Other Animals of Importance Considered Especially from an Economic Point of View. Printed by the Superintendent, Government Press, Madras. 565 pp.

Foott, W.H., 1977. Biology of the corn leaf aphid, Rhopalosiphum maidis (Homoptera: Aphididae), in southwesteren Ontario. The Canadian Entomologist 109 (8), 1129–1137.

Ghauri, M.S.K., 1971. Revision of the genus *Nephotettix* Matsumura (Homoptera: Cicadelloidea: Euscelidae) based on the type material. Bulletin of Entomological Research 60 (3), 481–512.

Ghosh, L.K., 2008. Handbook on Hemipteran Pest in India. Director, Zoological Survey of India, Kolkata, pp. 333–334.

Goodman, A., 1953. A Predator on *Creontiades pallidus* Ramb. Nature 171, 886.

Hampson, 1901. Catalogue of Lepidoptera Phalaenae in the British Museum by British Museum (Natural History). Dept. of Zoology. Hampson, George Francis, Sir, 1860–1936.

Hannikainen, M., 2011. Demise of Classic Maya Civilization a Theoretical Approach. (Bachelor Paper) School of Archaeology/Department of Culture, Energy and Environment, Gotland University (Uppsala University), Sweden. http://www.diva-portal.se/smash/get/diva2:431886/FULLTEXT01.pdf.

Hennig, W., 1965. Vorarbeiten zu einen phylogenetischen System der Muscidae (Diptera: Cyclorrhapha). Stuttgarter Beiträge zur Naturkunde, Serie A (Biologie) 141, 1–100.

Hunter, 1901. The Aphididae of North America. Bulletin of the Iowa Agricultural Experiment Station 60, 80.

Indian Museum Notes, vol. 3, 1900. Office of the Superintendent of Government Printing, Calcutta, India, p. 65.

Jotwani, M.G., Verma, K.K., Young, W.R., 1969. Observations on shoot fly, *Atherigona* spp. damaging different minor millet. Indian Journal of Entomology 31 (3), 291–293.

Kindler, S.D., Dalrymple, R.L., 1999. Relative susceptibility of cereals and pasture grasses to the yellow sugarcane aphid (Homoptera: Aphididae). Journal of Agricultural and Urban Entomology 16, 113–122.

Konstantinov, A.S., Baselga, A., Grebennikov, V.V., Prena, J., Lingafelter, S.W., 2011. Revision of the Palearctic Chaetocnema Species (Coleoptera: Chrysomelidae: Galerucinae: Alticini). Pensoft Publishers, Geo Milev Str. 13a, Sofia 1111, Bulgaria, pp. P69–P73.

Kfir, R., Overholt, W.A., Khan, Z.R., Polaszek, A., 2002. Biology and management of economically important lepidopteran cereal stem borers in Africa. Annual Review Entomology 47, 701–703.

Kruger, M., van den Berg, J., Du Plessis, H., 2007. A survey of sorghum panicle-feeding Hemiptera in South Africa. SAT eJournal. 3 (1). http://ejournal.icrisat.org/mpii/.

Leonard, D.E., 1966. Biosystematics of the "Leucopterus Complex" of the genus *Blissus* (Heteroptera: Lygaeidae). Bulletin/Connecticut Agricultural Experiment Station 677, 1–47.

Maes, K.V.N., 1995. A comparative morphological study of the adult Crambidae (Lepidoptera, Pyraloidea). Bulletin et Annales de la Société Royale Entomologique de Belgique 131, 383–434.

Maes, K.V.N., 1998. Pyraloidea: Crambidae, Pyralidae. In: Polaszek, A. (Ed.), African Cereal Stem Borers: Economic Importance, Taxonomy, Natural Enemies and Control. CABI, Wallingford, UK, pp. 87–98.

Malloch, J.R., 1923. Exotic Muscidae (Diptera) – X. Annals and Magazine of Natural History 12, 177–194.

Malloch, J.R., 1924. Exotic Muscaridae (Diptera). – XII. Annals and Magazine of Natural History 13 (9), 409–424.

Mayr, E., Ashlock, P.D., 1991. Principles of Systematic Zoology. McGraw-Hill, Singapore. xx, 475 p.

Medina-Gaud, S., Martorell, L.F., Robles, R.B., 1965. Notes on the biology and control of the yellow aphid of sugarcane, *Sipha flava* (Forbes) in Puerto Rico. In: Proceedings of the 12th Congress of the International Society of Sugarcane Technology, San Juan, Puerto Rico, March 28–April, 10, pp. 1307–1320.

Minet, J., 1981. Les Pyraloidea et leurs principales divisions systématiques. Bulletin de la Société Entomologique de France 86, 262–280.

Minet, J., 1983. Étude morphologique et phylogénétique des organes tympaniques des Pyraloidea. 1. Généralités et homologies (Lepidoptera Glossata). Annales de la Société Entomologique de France 19 (2), 175–207.

Moiz, S.A., Naqvi, K.M., 1968. Studies on sorghum stem fly *Atherigona varia* var. *soccata* Rondani (Anthomyiidae: Diptera). Agriculture Pakistan 19, 161–164.

Moyal, P., 2006. History of the systematics of the *Sesamia sensu lato* group of African noctuid stem borers of monocotyledonous plants (Lepidoptera). Annales de la Société Entomologique de France 42 (3/4), 285–291.

Moyal, P., Tokro, P., Bayram, A., Savopoulou-Soultani, M., Conti, E., Eizaguirre, M., Rü, B.Le, Avand-Faghih, A., Frérot, B., Andreadis, S., 2011. Origin and taxonomic status of the Palearctic population of the stem borer *Sesamia nonagrioides* (Lefèbvre) (Lepidoptera: Noctuidae). Biological Journal of the Linnean Society 103 (4), 904–922.

Nault, L.R., 1983. Origins of leafhopper vectors of maize pathogens in Mesoamerica. pp. 75–82. In: Gordon, D.T., Knoke, J.K., Nault, L.R., Ritter, R.M. (Eds.), Proceedings International Maize Virus Disease Colloquium and Workshop, 2–6 August 1982. The Ohio State University. Ohio Agricultural Research and Development Center, Wooster, Ohio. 266 pp.

Odhiambo, T.R., 1958. Notes on the East African Miridae (Hemiptera). V: New species of *Eurystylus* Stål from Uganda. Annals and Magazine of Natural History 13 (1), 257–281.

De Prins, J., De Prins, W., 2013. Afromoths, Online Database of Afrotropical Moth Species (Lepidoptera). World Wide Web Electronic Publication. www.afromoths.net.

Pape, T., Thompson, F.C. (Eds.), 2013. Systema Dipterorum (version 2.0, Jan 2011). In: Species 2000 and ITIS Catalogue of Life, 11th March 2013. Species 2000, Reading, UK. Roskov, Y., Kunze, T., Paglinawan, L., Orrell, T., Nicolson, D., Culham, A., Bailly, N., Kirk, P., Bourgoin, T., Baillargeon, G., Hernandez, F., De Wever, A. (Eds.). Digital resource at: www.catalogueoflife.org/col/.

Pérez Hidalgo, N., Martínez-Torres, D., Collantes-Alegre, J.M., Villalobos Muller, W., Nieto Nafría, J.M., 2012. A new species of *Rhopalosiphum* (Hemiptera: Aphididae) on *Chusquea tomentosa* (Poaceae: Bambusoideae) from Costa Rica. Zookeys 166, 59–73.

Pont, A.C., 1972. The oriental species of *Atherigona* Rondani. In: Jotwani, M.G., Young, W.R. (Eds.), Control of Sorghum Shoot Fly. Oxford and IBH Publishing, New Delhi, India, pp. 27–104.

Pont, A.C., Magpayo, F.R., 1995. Muscid shoot-flies of the Philippine islands (Diptera: Muscidae, genus *Atherigona* Rondani). Bulletin of Entomological Research (Suppl. 3), 123 750 figs.

Prior, R.N.B., 1976. Keys to the British species of *Metopolophium* (Aphididae) with one new species. Systematic Entomology 1 (4), 271–279.

Ramachandra Rao, Y., 1924. The genitalia of certain Anthomyiid flies (*Atherigona* spp.). In: Report of the Proceedings of 5th Entomology Meeting, 1923, Pusa, Bihar, India, pp. 330–335.

Rashmi, G., Devinder, S., 2013. Taxonomic notes on five species of the genus *Cletus* Stal (Heteroptera: Coreidae) from northern India with particular reference to their female genitalia. Journal of Entomology and Zoology Studies 1 (6), 44–51.

Royer, T.A., 2015. Management of Insect and Mite Pests in Sorghum, CR-7170. Oklahoma State University Cooperative Extension Service, Stillwater, OK.

Seshu Reddy, K.V., Omolo, E.O., 1985. Sorghum insect pest situation in eastern Africa. In: Proceedings of the International Sorghum Entomology Workshop, 15–21 July 1984. Texas A&M University and ICRISAT, pp. 31–36.

Sharma, H.C., Davies, J.C., 1988. Insect and Other Animal Pests of Millets. Sorghum and Millets Information Center, Patancheru 502 324, Andhra Pradesh, India: International Crops Research Institute for the Semi-Arid Tropics, 142 pp.

Sharma, H.C., Lopez, V.F., 1990. Mechanisms of resistance in sorghum to head bug, *Calocoris angustatus*. Entomologia Experimentalis et Applicata 57 (3), 285–294.

Simpson, G.G., 1961. Principles of Animal Taxonomy. Columbia University Press, New York.

Singh, B.U., Seetharama, N., 2008. Hose plant interactions of the planthopper, *Peregrinus maidis* Ashm. (Homoptera: Delphacidae) in maize and sorghum agroecosystems. Arthropod-Plant Interactions 2, 163–196.

Singh, B.U., Padmaja, P.G., Seetharama, N., 2004. Biology and management of the sugarcane aphid, *Melanaphis sacchari* (Zehntner) (Homoptera: Aphididae), in sorghum: a review. Crop Protection 23, 739–755.

Solis, M.A., 2007. Phylogenetic studies and modern classification of the Pyraloidea (Lepidoptera). Revista Colombiana de Entomología 33 (1), 1–9.

Starks, K.J., Burton, R.L., 1977. Preventing Greenbug Outbreaks Report No. 309. USDA Sci. Educ. Admin. Leaflet, Washington, DC.

Stroyan, H.L.G., 1984. *Aphids—Pterocommatinae and Aphidinae* (Aphidini). Handbooks for the Identification of British Insects 2, Part 6. Royal Entomological Society of London, London, p. 232.

Sujatha, N.K., Pape, T., Pont, A.C., Wiegmann, B.M., Meier, R., 2008. The Muscoidea (Diptera: Calyptratae) are paraphyletic: evidence from four mitochondrial and four nuclear genes. Molecular Phylogenetics and Evolution 29 (49(2)), 639–652.

Sutrisno, H., 2012. Molecular Phylogeny of Indonesian Armyworm *Mythimna* Guenée (Lepidoptera: Noctuidae: Hadeninae) Based on CO I Gene Sequences. HAYATI Journal of Biosciences 19 (2), 65–72.

Swinhoe, C., 1885. On the Lepidoptera of Bombay and the Deccan. Heterocera. Proceedings of the Zoological Society of London 1885 (2), 287–307.

Tams, W.H.T., Bowden, J., 1953. A revision of the African species of Sesamia Guenée and related genera (Agrotidae-Lepidoptera). Bulletin of Entomological Research 43, 645–678.

Webb, M.D., 1987. Species recognition in *Cicadulina* leafhoppers (Hemiptera: Cicadellidae), vectors of pathogens of Graminep. Bulletin of Entomological Research 77 (4), 683–712.

Webster, J.A., 1995. Economic Impact of the Greenbug in the Western United States, 1992–1993. Vol. 155. Great Plains Agricultural Council, Stillwater, OK.

Webster, F.M., Phillips, W.J., 1912. The Spring Grain-Aphis or "Greenbug". U.S. Department of Agriculture, Burnet Entomology Bulletin, p. 133.

Wenn, C.T., 1964. A study on the millet stem fly, *Atherigona biseta* Karl. Acta Phytophylacica Sinica 3, 41–48.

Wu, J.T., 1981. On an unrecorded sugarcane pest, *Sesamia uniformis* Dudgeon (Noctuidae: Lepidoptera). Journal of South China Agricultural College 2 (3), 14–20.

Youm, O., Harris, K.M., Nwanze, K.F., 1996. Coniesta ignefusalis (Hompson), the millet stem borer:a handbook of information. In En. Summaries in En, Fr, Es. Information Bulletin no. 46. Patancheru 502324, Andhra Pradesh, India: International Crops Research Institute for the Semi-Arid Tropics. 60 pp. [Part 1: Review pp. 118; Part·2: Annotated Bibliography pp. 19–52] ISBN 92-9066-253-0.

Young, W.R., Teetes, G.L., 1977. Sorghum entomology. Annual Review of Entomology 22, 193–218.

BIOLOGY, BEHAVIOR, AND ECOLOGY

MORPHOLOGY AND ANATOMY
SHOOT FLIES

All the relevant morphological details of the Muscidae are discussed in this chapter for a comparative overall understanding of *Atherigona*. In general, the eggs of muscid flies are morphologically equipped with specialized chorionic structures. Such chorionic adaptations provide plastronic effects enabling the eggs to withstand wet or fluidic ovipositional environments.

Based on hatching pleats and chorionic structure, muscid eggs are classified into two types (Skidmore, 1985).

1. *Musca* type: The *Musca* type has slightly raised carinae forming hatching pleats along which the chorion splits during larval hatch. The chorion is entirely formed by three layers, viz., outer, median, and inner meshwork layers. It lacks an external chorionic envelope. As an exception, some species of *Musca* and *Morellia* possess an external chorionic envelope. A median stripe produced dorsally develops into respiratory horn in the eggs of those species.
2. *Phaonia* type: In the *Phaonia* type the hatching pleats are foliaceous. In contrast to the *Musca* type, the dorsal strip alone is covered with an outer meshwork layer. Flies of the Mydaeinae and Coenosiinae lay eggs with hatching pleats produced anteriorly to form paired respiratory horns.

Atherigona **eggs are of the** *Phaonia* **type. Eggs are elongate, ovate, slightly concave dorsally, and convex ventrally.**

The head of larvae is longitudinally bilobed, the oral cavity opens between the lobes. The sensorial papillae and the oral grooves spread out fanwise from the oral opening. The head is called a pseudo-cephalon because the larvae lack a distinct head capsule. In *Atherigona reversura*, each lobe of the bilobate is equipped with an antennal complex, maxillary palpus, ventral organ, and labial organ, and surrounding the functional mouth opening is a facial mask composed of cirri arranged in short rows and oral ridges (Grzywacz et al., 2013).

Muscid larvae exhibit polymorphism in having differences in the development of spiracles at successive instars. Usually all the muscid larvae look alike except in size and in the development of spiracles. The larvae can be broadly classified as follows (Ferrar, 1979; Skidmore, 1985) based on morphology:

1. Monomorphic: First and second instars are passed inside the egg; a sieving mechanism occurs in the oral cavity; anal spiracles are minute.
2. Dimorphic: First-instar larva passes its lifetime inside the egg; a sieving mechanism occurs in the oral cavity.

Insect Pests of Millets. http://dx.doi.org/10.1016/B978-0-12-804243-4.00003-3

3. Trimorphic: All the instars live outside the egg and each differs in morphological details; anal spiracles are massive.

Some predatory muscid larvae (e.g., *Phaonia*) in their third instar possess complex oral bars and associated ribbons. The oral bars in some species (e.g., *Atherigona* s.str.) become massively enlarged and taken over the function of the mouth hooks. This clearly points to the fact that phytophagy in Muscidae evolved from predaceous antecedents (Skidmore, 1985). In the genus *Atherigona* itself there are differences in larval feeding habits. The subgenus *Achritochaeta* shows predatory behavior and lives on decomposed materials, whereas *Atherigona* s.str. is phytophagous. However, there are some logical doubts on the feeding habits of the species of *Atherigona* s.str. as being considered true phytophagy.

Postcephalic segments in muscid larvae carry imaginal leg and wing discs (Ranade, 1965). There are eight visible abdominal segments and the 8th segment fused with the 9th and 10th forms the caudal segment. The anal spiracles of the mature larvae are taxonomically important (Skidmore, 1985).

There are five types of larvae recognized in the Muscidae based on larval biology and morphology (Skidmore, 1985):

1. Trimorphic saprophage: The anal spiracles are massive. Examples are Eginiinae, Muscini, Stomoxyinae, and some Reinwardtiinae.
2. Trimorphic obligative phytophage: The oral bar is massively enlarged and often toothed beneath. An example is *Atherigona* s.str.
   The sorghum shoot fly, *Atherigona soccata* posses ventrally toothed oral bar.
3. Trimorphic facultative carnivore: The anal spiracles are large to very large. Examples are Reinwardtiinae, Achanthipterinae, Azeliinae, Mesembrinini, and *Acritochaeta*.
4. Dimorphic obligative carnivore: The anal spiracles very small. Examples are Mydaeinae, Coenosiinae, Phaoniinae, and some Azeliinae.
5. Monomorphic obligative carnivore: The smallest anal spiracles can be seen in this group.

Shoot flies belonging to *Atherigona* s.str. fall in the trimorphic obligate phytophagous group, whereas the subgenus *Achritochaeta* of the genus *Atherigona* are trimorphic facultative carnivores and phaoniines exhibit dimorphic obligate carnivory. For example, larvae of *Atherigona soccata* (subgenus *Atherigona* s.str.) feed on decomposed tissues of the sorghum plant. The pepper fruit fly, *Atherigona orientalis*, belonging to subgenus *Achritochaeta*, is polyphagous. Larvae of *A. orientalis* feed and develop on decaying plant material, feces, and carrion and also feed on the larvae of *Dacus* spp. (Skidmore, 1985) and *Bactrocera* spp. (Uchida et al., 2006).

MORPHOLOGY OF *ATHERIGONA SOCCATA*

Egg Morphology

The egg is 1.2 mm in length, white, cigar shaped, and elongated (Fig. 3.1). The egg is trilobed longitudinally with the leathery outer shell having a netlike ultrastructure.

Larval Morphology

The larvae are trimorphic. The third-instar larva has three thoracic (t1–t3) and eight abdominal (a1–a7) segments (Fig. 3.2). A pair of anterior spiracles (Fig. 3.3) is located laterally on the first thoracic segment. The anterior spiracle is a conspicuous digit like eight-lobed structure. The

FIGURE 3.1

Egg of *Atherigona soccata*.

Courtesy of A. Kalaisekar.

FIGURE 3.2

Third-instar larva of *Atherigona soccata*. *a1–a7*, Abdominal segments; *t1–t2*, thoracic segments.

Courtesy of A. Kalaisekar.

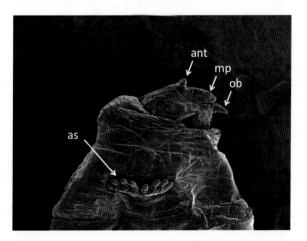

FIGURE 3.3

Anterior portion of third-instar larva of *Atherigona soccata* showing anterior spiracles (scanning electron microscope image). *as*, anterior spiracles; *ant*, antenna; *mp*, maxillary palpus; *ob*, oral bar.

Courtesy of A. Kalaisekar.

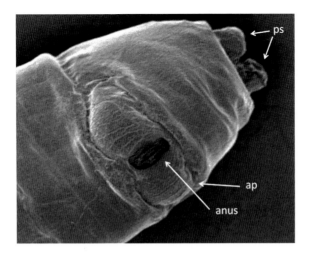

FIGURE 3.4

Posterior portion of third-instar larva of *Atherigona soccata* showing anal plate and posterior spiracles (scanning electron microscope image). *ap*, anal plate; *ps*, posterior spiracles.

Courtesy of A. Kalaisekar.

pseudocephalon is symmetrically bilobed and each lobe has an antenna, a maxillary palpus, and a large oral bar. The posterior region (Fig. 3.4) has an anus surrounded by a raised anal plate. A pair of posterior spiracles is located terminally.

Cephalopharyngeal Sclerites in Larva

An extensive and highly sclerotized cephalopharyngeal skeletal system can be seen internally at the anterior end of the larva. The structure of the cephalopharyngeal skeleton greatly varies with instars (Raina, 1981). A pair of mandibular sclerites extends out to produce a mouth hook each. The mouth hooks are sharp in the first instar and then become short and a little blunt in the second instar. In the third instar, the mandibular sclerite becomes a ventrally toothed mandible. A hypostomal sclerite connects the mandibular sclerite with the pharyngeal sclerite. The pharyngeal sclerite is connected to a massive forklike structure with the cornus (dorsal cornu and ventral cornu) of the fork directed backward. Cephalopharyngeal sclerites are taxonomically important structures.

Gut and Salivary Glands

The digestive canal of the larva of *A. soccata* is a simple tube, without much differentiation. The gut tube shows a short stomodeum, a long mesenteron, and a proctodeum. A pair of salivary glands beneath the mesenteron opens into the preoral cavity (Dabrowski and Patel, 1981).

Pupal Morphology (in Latin, *Pupa*, Meaning Doll)

In general the pupal stage of muscid flies is called the puparium because the contracted skin of the final-instar larva forms the pupa. After puparial formation most of the external features of the last larval instar are visible in a distorted state and the mouthparts are seen in the thoracic segments. Pupal horns protrude through the dorsolateral walls of the first abdominal segment and help in pupal respiration (Skidmore, 1985). The typical puparium of *Atherigona* is truncated, barrel shaped, and thick orange to dark reddish or dark brown in color. In *A. reversura* the anal spiracular area is surrounded with a fold (Grzywacz et al., 2013). The truncated barrel-shaped pupa retains the anterior and posterior spiracles at the top and base, respectively (Fig. 3.5).

FIGURE 3.5

Puparium of *Atherigona soccata. as*, anterior spiracles; *ps*, posterior spiracles.

Courtesy of A. Kalaisekar.

Reproductive System

In the male reproductive system, an externally visible trifoliate process is unique to the subgenus *Atherigona* s.str. The structure of the trifoliate process is of great importance in the species delimitations in the subgenus. The trifoliate process assists in clasping the female during copulation.

The internal reproductive system in the female shoot fly is similar to that of other muscid flies. It consists of a pair of ovaries, the lateral oviducts, the median oviduct, a pair of accessory glands, and three spermathecae (Unnithan, 1981). The accessory glands are kidney shaped. The spermathecae are black and oval with a whitish apical soft cap. The ovaries contain polytrophic ovarioles with 15 trophocytes for nutritional support of each follicle. On average about 34 ovarioles are distributed unequally between the right and the left ovaries in *A. soccata* (Unnithan, 1981).

Egg maturation takes place synchronously in all the ovarioles in young females, with only one egg maturing at a time in each ovariole. Each developing follicle contains 15 nutritive cells (trophocytes) and 1 oocyte. In newly emerged flies the follicles are not differentiated. Absence of mating delays or prevents ovipositioning in females (Unnithan, 1981) and fecundity is greatly influenced by the host plant (Ogwaro, 1978).

STEM BORERS

Egg

Chilo spp. and *Maliarpha* spp. do not cover the egg with a tuft of hairs. *Chilo partellus* lays eggs in a herringbone pattern usually on the upper surface of the leaf along the midrib (Fig. 3.6). Eggs of *Maliarpha separatella* are usually protected in leaf folds around each egg mass. *Busseola fusca* lays eggs in a mass, not covered with a tuft of hairs, and individual eggs are hemispherical with crenulations (radial vertical ridges on the egg shell). *Sesamia* spp. also lay eggs in a mass and the eggs are generally almost elliptical with a flat base. No special structure is appended with eggs. *Sesamia inferens* eggs are beadlike. *Sesamia grisescens* eggs are barrel shaped with longitudinal striations. Varying numbers of micropyles surrounded by petaloidal formations are present. A series of irregular polygonal plaques is seen with numbers varying with species (Pucci and Forcina, 1984).

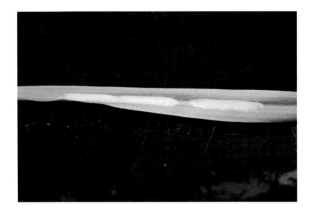

FIGURE 3.6

Egg mass of *Chilo partellus* in herringbone pattern on the upper surface of the leaf along the midrib.

Courtesy of A. Kalaisekar.

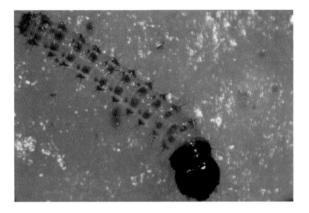

FIGURE 3.7

First-instar larva of *Chilo partellus* with prominent radiating setae.

Courtesy of A. Kalaisekar.

Larva

The larva of *C. partellus* possesses two longitudinal dorsal series of spots between two dorsolateral series of spots. Such spots earned *C. partellus* its name, spotted stem borer. Each spot has two or three radiating setae, which are more prominent in first-instar larvae (Fig. 3.7). The head and the first thoracic segment are dark brown in *C. partellus*. The crochets are arranged in a circular fashion on the prolegs in *Chilo* spp. The arrangement of crochets is semicircular in *Sesamia* spp. Larvae of *B. fusca* and *S. inferens* look similar and can be differentiated by color. *B. fusca* larvae have a dark brown head and yellowish brown prothorax.

The larval and adult antennae of *C. partellus* possess hygro/thermosensitive sensilla sensitive to minute changes in temperature. The grooved basiconic sensilla in the male antennae perceive female sex pheromones. The smooth-surfaced basiconica and coeloconica sensilla are sensitive to the host-plant odors. Larval maxillary styloconica sensilla can detect differences between saps from susceptible and resistant host plants (Waladde et al., 1990).

Pupa

Pupae of all the lepidopteran borers are obtect. The pupae have apical protruberances and indented lines around in segments 5–7 in *Chilo polychrysus*. Cornicles bearing frontal tubercles are raised conspicuously. Pupae of *Sesamia calamistis* possess two dorsal spines as cremaster (= to hang, in Latin), a series of small hooks or a single larger hook used to attach the posterior end of a pupa to a twig or other structure.

SHOOT BUG

Mouthparts

The shoot bug *Peregrinus maidis* shares the common structure of the mouthparts in Fulgoroidea. There are three T-shaped dorsal, middle, and ventral locks internally formed between the maxillae. The food and salivary canals are usually formed by maxillae. The mandibles are contiguously connected with maxillae by an external wall (Brożek et al., 2006).

Alimentary Canal

The alimentary canal is a simple fulgoroid type. There is no filter chamber. The cibarial diaphragm is connected through the esophagus to the midgut. The saclike diverticulum forms an extension from the midgut. Two pairs of Malpighian tubules arise at the end of the midgut. The rectal sac and rectum form the hindgut (Tsai and Perrier, 1993).

Salivary Glands

The structure and function of the salivary glands in *P. maidis* are important in terms of this bug being the vector of the two most dreaded plant viruses, namely maize mosaic virus (MMV) and maize stripe virus (MStrV). Each salivary gland consists of an accessory and a principal gland. Muscle cells or myofibrils in the salivary glands are not known to occur in insects except in *P. maidis*, *Myzus persicae*, and the aster leafhopper, *Macrosteles fascifrons* (Cicadellidae) (Ammar, 1986). Generally, the salivary secretions are transported from secretory cells to the intracellular canaliculi (= fine tubular channel; singular: canaliculus) through the process of pinocytosis and then to intercellular ducts, which eventually open into the salivary canal. In addition to pinocytosis, the myofibrils in the salivary glands assist in transportation of secretory materials by way of pulsations (Ponsen, 1977). The bug transmits MMV and MStrV from infected plants to healthy ones. The sap containing the virus from the diseased plant goes into the gut of the bug and then invades the hemocoel. The virus then gets entry into the salivary glands and mixes with the saliva to ultimately enter the healthy plant when the bug feeds.

Reproductive System

The male reproductive system has two testes, each consisting of three or four follicles and with large accessory glands (Tsai, 1996). Females possess two large ovaries, each with around 16 ovarioles (Tsai and Perrier, 1993). As in other insects, during copulation in *P. maidis*, the spermatophore is deposited in the bursa copulatrix. The spermatophore is further pushed into the spermatheca through peristaltic contractions. The spermathecal glands produce enzymes to digest the protein coat of the spermatophore and also provide nutrients for the spermatozoa.

The morphology of the spermatozoa of *P. maidis* is clearly understood in great detail (Herold and Munz, 1967). A spermatozoon is threadlike, approximately 650 μm long and 1 μm wide. The spermatozoa consist of two mitochondrion derivatives, a central body, the axial filament complex, and two wing-shaped bodies. Each mitochondrion derivative shows a peripheral part formed by cristae and a crystalline array of the inner part. The cristae are approximately 70 Å wide and the inner part approximately 100 Å in diameter. The head has an elongated nucleus (Herold and Munz, 1967).

APHIDS

The morphology of aphids is important in recognizing their species identity. Owing to the complexities in life cycles, the taxonomic identification of aphids is often a difficult task. In the aphids that undergo obligatory annual sexual cycles (holocyclic), the smallest taxonomic units are called species. In the case of aphids that reproduce through parthenogenesis (anholocyclic) the smallest units are called clonal lineages (Blackman et al., 1989). Aphids are vectors of several viral diseases of plants.

Mouth parts

Aphids have piercing and sucking mouthparts like true bugs. Mandibles and maxillae are sheathed within a modified labium, called the rostrum, and the entire setup of this structure is known as the stylet. Stylets are akin to a hypodermic needle. A stylet is capable of piercing plant tissue and reaching to the phloem of the plant. The phloem is under heavy pressure and aphids carefully negotiate and regulate the sap flow into the stomach to avoid exploding due to phloem-sap pressure.

MIDGE

Larvae of the sorghum midge, *Stenodiplosis sorghicola*, possess a prominent labroclypeal plate with a flattened clypeus and a falcate labrum. The mandible is crescent shaped with a single well-defined tooth at the end. Maxillae are large and elongate. The labium tapers from the base. These mouthparts are adapted for feeding on plant juice (Petralia et al., 1979).

PHYSIOLOGY AND NUTRITION
DIGESTIVE PHYSIOLOGY

Unlike mammalian digestion, which takes place under acidic conditions, the digestion in insects occurs at neutral or basic pH. In the larvae of *Drosophila melanogaster* that feed on decaying vegetable matter, microbes, especially *Lactobacillus* and *Acetobacter*, enter the gut through contaminated food and assist in digestion (Bruno and Irene, 2013). Similarly, there is a need to investigate if the larvae of *A. soccata* also acquire such microbial colonies through decaying plant tissues on which they feed.

METABOLIC PHYSIOLOGY

In general, fat-body physiology is very important because lipid (a component of the fat body) metabolism meets the energy needs of the extended nonfeeding life periods in insects. There are at least four kinds of specialized cells found in various insects, namely adipocytes, urocytes, mycetocytes, and oenocytes (Lockey, 1988; Estela and Jose, 2010). Adipocytes store fat bodies in the form of lipids and glycogen. In insects, fat bodies are bathed in hemocoel facilitating the supply of energy through release of trehalose during an intense need for energy, e.g., in flight. Urocytes store the excretory material (urate) in insects such as cockroaches and grasshoppers (Estela and Jose, 2010). Urocytes are absent in lepidopterans, and adipocytes accumulate urate, and such cells disappear during the adult eclosion (Willott et al., 1988). Mycetocytes harbor microbial symbionts. Oenocytes are associated with the synthesis of cuticular lipids, proteins, and hydrocarbons (Lockey, 1988; Estela and Jose, 2010).

Lipase hydrolyzes triglycerides to diglycerides and fatty acids. The lipase activity in the digestive physiology of *C. partellus* fluctuates with the age of the pupa. Lipase activity increases in the 19-day-old larva and 4-day-old pupa and decreases thereafter (Pol and Sakate, 2002; Sakate and Pol, 2002).

Male accessory reproductive glands contribute proteins to the spermatophore nutrition, and these proteins can be of either intraglandular or extraglandular origin (Ismail et al., 1997).

NUTRITION

The size of the food plant influences the survival and rate of larval development in *A. soccata* (Ogwaro and Kokwaro, 1981a,b). Carbohydrate supplements as adult food accelerate vitellogenesis and shorten the preovipositioning period. Carbohydrate deficiency causes oocyte resorption. Water and protein supplements increase resorption of oocytes and the preovipositioning period while reducing the number of eggs (Unnithan, 1981). Ovipositioning in adult females increases when yeasts are supplemented with sugar and water (Raina, 1982), and such food materials increase the adult life span too (Meksongsee et al., 1981). Longevity and fecundity of *A. soccata* are mainly affected by the adult's food (Raina, 1982). The absence of protein in the diet of female *Delia antiqua* resulted in lower search attraction and reproductive competency of fecundity (McDonald and Borden, 1996). In general, protein deficiency at the adult stage seriously affects the reproductive success of many muscoid dipterans (Jones et al., 1992).

The quantity of cyanide and phenolic compounds in sorghum influences the feeding behavior of *P. maidis*. Young sorghum plants generally produce more cyanides. Feeding of *P. maidis* on young plants is reduced because of the presence of mixtures of phenolic acids and their esters. This is primarily due to high phenolic acid concentrations, which reduce the ability of *P. maidis* to locate the phloem tissues (Fisk, 1980).

REPRODUCTIVE BEHAVIOR AND LIFE HISTORY
SHOOT FLIES

The distribution pattern of eggs in the sorghum shoot fly, *A. soccata*, is random or slightly aggregated rather than regular, irrespective of the presence of other eggs already laid (Delobel, 1981), and similarly, the egg distribution of *Delia arambourgi* (Anthomyidae), a shoot fly infesting tef, is patchy, and sufficient eggs are laid to permit attacks on all available shoots (Bullock, 1970). Host preferences of shoot flies are driven by the selectivity of the ovipositing female rather than the dietary requirements of the larvae (Nye, 1959). But in butterflies, oviposition site selection is an act of parental care and it is an important life history trait with significant consequences for offspring development and success (Hans van and Sofie, 2010). Therefore, eggs are not deposited randomly by butterflies, as females select appropriate oviposition sites so as to afford protection to the eggs and ensure food resources for the hatching offspring (Gullan and Cranston, 2010). In *A. soccata*, usually one or two eggs (average 1.2 eggs/plant) are laid by a female on a plant and egg distribution follows a Poisson distribution (Raina, 1982). Such an act of egg moderation by *A. soccata* could possibly be a tactic to overcome sibling conflict during the larval feeding period inside the shoot of the plant seedling. Therefore, the egg distribution needs to be studied with a view to see if it amounts to an act of parental care by the female fly. Egg size is a critical aspect of an insect's reproductive strategy that would help in food resource allocation to offspring (Michael, 1994). Frit flies, *Oscinella frit*, lay almost all their eggs in a group behind a single coleoptile when in captivity, whereas under conditions of an active predatory population, 60% of all eggs are laid singly (Margaret, 1969). The larva of *O. frit* takes more than one host plant seedling for its development. But in the case of *A. soccata*, larval development is completed within a single seedling of the host plant and in *A.soccata,* all the eggs are laid behind the single coleoptile of even 3-4 leaved seedling under captivity like in *O.frit* (Kalaisekar et al., 2013).

Plant height plays an important role in oviposition preference and larval survival in *A. soccata*. Survival of the first-instar larvae is influenced by the resistance to penetration of the leaf sheaths (Delobel, 1982; Kalaisekar et al., 2013). The females select oviposition sites through a succession of probing movements of the anterior tarsus and of the ovipositor (Ogwaro, 1978). The diurnal fly is more active in the morning and evening; the female lays cigar-shaped white eggs singly on the underside of the leaves. Ambient temperature affects the development rate of the immature stages (Raina, 1981). Egg maturation and fecundity usually remain unaffected by multiple matings or by the presence of males. Females generally mate only once and this ensures fertilization of all eggs laid (Unnithan, 1981). Lack of mating reduces fecundity and affects the rate of oviposition by females, but does not affect longevity. Females may withhold eggs because of a lack of suitable oviposition substrate. Unsuitable grass hosts reduce the fecundity but do not affect the longevity of adult flies (Raina, 1982). Fecundity varies with the age of the fly (Nwanze et al., 1998). Low temperatures prolong the adult life span and preoviposition and oviposition periods and reduce daily egg production, mean fecundity, and fertility (Unnithan et al., 1985). The survival of the egg is at stake with decreasing humidity levels (Delobel, 1983). Oviposition site varies in some species of *Atherigona*, for example, *Atherigona naqvii* lays more eggs on the soil around the plant than on the stalk and foliage (Sarup and Panwar, 1987). Adult flies of *Delia flavibasis*, a shoot pest of tef, are usually found on newly ploughed moist soil and newly emerging seedlings. The flies are more active during late morning and late afternoon (Goftishu et al., 2004a,b).

Oviposition starts at the third leaf stage (1 week after germination) of the plant. The largest numbers of eggs are laid on the third and fourth leaves (Ponnaiya, 1951). The number of eggs laid by a shoot fly ranges from 20 to 25 (Kundu and Kishore, 1970). The egg hatches within 48–72 h. Development time from egg to adult is 14–28 days. There are several bursts of ovipositioning that occur over a period of 4–5 weeks after a preovipositioning period of 3–4 days. The mean number of eggs to females is 62.8, and the mean preovipositioning period is 4.7 days. Mean longevity is 32.6 days for females, 25.4 days for males. A freshly laid egg measures 1.1–1.2 mm. Full-grown larvae are 6.2–9 mm in length, 0.8–1.5 mm in width. Pupae are 0.9–16 mm wide, 3.2–4.5 mm long. Adult males are 3–4.3 mm long, 0.6–1.1 mm wide. Females are 3.3–4.6 mm long, 0.8–1.2 mm wide. After 1 or 2 days of mating, copulation lasts for 5–20 min. The preovipositioning period is 2–4 days. Only one first-instar larva can survive and develop in a single stem (Meksongsee et al., 1978; Unnithan, 1981; Delobel, 1982; Raina, 1982). Twenty to twenty-five eggs are laid in a period of 12–24 days at one or two eggs per day (Kunder and Kishore, 1970); and there are reports of as may as 235 eggs laid by a single female fly (Meksongsee et al., 1978).

The egg period lasts for about 61 h. The first-instar larva enters the plant through the gap between the leaf sheath and the growing central shoot in a period of 30 min. Longer exposure to the ambient weather prevailing on the leaf or plant surface causes larval death due to desiccation. Moisture on the leaf surface also plays an important role in survival and movement of the larvae after hatching. Larvae spending more time on the plant surface run the risk of losing leaf surface wetness as well as getting exposed to the vagaries of air humidity changes. The egg is laid on under the surface of the leaf and is always positioned upright. The larva hatching from the egg takes a U turn and crawls down for some distance to turn toward the leaf margin and crosses over onto the upper surface of the leaf blade. Upon reaching the upper leaf surface, the larva crawls down toward the leaf base and enters into the plant shoot between the leaf sheath and the growing shoot whorl. From this moment of larval entry into the plant until total extrication of the adult fly during eclosion, *A. soccata* passes its life endophytically. The first symptoms of withering of the central shoot after larval entry into the plant can be noticed in 34 h. The duration of the first-instar larva is 35 h. From second instar to fully matured last instar takes 127 h and the larva feeds on the inner rotten shoot tissues. By this time the plant shows the typical dried-up central growing shoot called a "deadheart."

The pupal duration is 124 h. The ptilinal force opens up the puparial lid, helping the subimago to gradually extricate itself from the puparium. The eclosion process takes 9.8 h. The posteclosion phase of *A. soccata* involves a series of well-defined activities, viz., wing expansion, tanning, grooming, and flight. From larval hatching to adult requires an average of 310 h (Fig. 3.8; Kalaisekar et al., 2013).

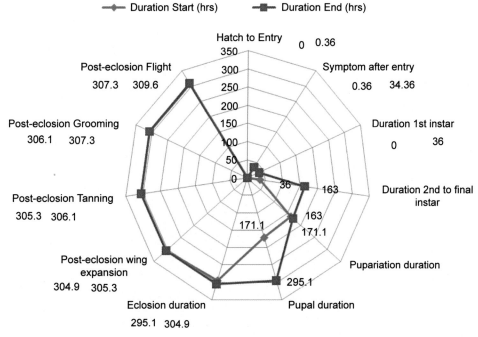

FIGURE 3.8

Life cycle of *Atherigona soccata* from larval hatching to adult flight, with start and end durations of various biological events depicted in sequence.

Courtesy of A. Kalaisekar.

STEM BORERS

C. partellus completes one or more generations per year, depending on the location and the host's availability (Muhammad and Underwood, 2004). Under favorable conditions such as warm low-altitude regions and with ample host availability, *Chilo* undergoes several cycles throughout the year. Larval diapause is observed at higher elevations or during dry seasons (Kfir et al., 2002). Generally adult emergence takes place late in the afternoon or early in the evening. Female moths prefer the whorl or funnel stage of the plants for oviposition. Eggs are oval shaped, flattened, and laid in clusters that hatch in 7–10 days (Ofomata et al., 2000). Mating takes place soon after emergence of the females. After a preovipositioning period of about 3–4 days, each female lays up to 200–600 scale like overlapping eggs in 10–80 separate batches on the undersides of leaves, mostly near the midribs.

The eggs of *Chilo* hatch in 4–10 days, usually between 0600 and 0800 h. Early instar larvae move into the whorl and feed on leaf surfaces deep inside the whorl. Later instars move down the stalk or bore into the stalk just above an internode. In older plants, larvae feed inside the internodal areas and make extensive galleries. Larvae sometimes feed on the developing earheads. Fully grown larvae are up to 25 mm long. Under warm conditions larval development takes about 15–20 days. Larvae pupate within the plant stalk. Prior to pupation the fully grown larva cuts out an exit hole to facilitate adult emergence from the plant (Fig. 3.9). Pupae are up to 15 mm long. The pupal period usually lasts for about 5–12 days. The total life cycle is completed in 25–50 days when climate and growing conditions are favorable.

Pupae of *Chilo* spp. and *Sesamia* spp. are without cocoons and those of *Mythimna separatella* are covered with whitish silken cocoons. *S. inferens* usually pupates inside the stem and rarely between the leaf sheath and the stem. The adult exit hole is cut out by the full-grown larva before pupation and the external opening of the exit hole is usually covered with silk.

C. partellus undergoes facultative diapause (Kfir, 1991) and spends the dry season in diapause in residues (Taley and Thakare, 1980). The diapause is induced by the dry condition of the host plant (Nye, 1960)

FIGURE 3.9

Adult larvae of *Chilo partellus* cutting exit hole before pupation.

Courtesy of A. Kalaisekar.

and the deterioration of the nutritive environment (Scheltes, 1978a,b). However, in the case of availability of wild grasses, actively feeding larval populations are also observed in addition to the diapausing larvae (Ngi-Song et al., 1996; Mbapila et al., 2002). In southern Africa, *C. partellus* undergoes diapause during the cold dry season (April–October) and there are overlapping generations of *C. partellus* (Kfir, 1988, 1998) observed in several areas, which favors infestations throughout the growing season.

The optimum range of temperature for the development of *C. partellus* lies between 20 and 30°C (Jalali and Singh, 2004). The developmental threshold temperature ranges between 10 and 40°C and the degree-days number required to complete a generation is 802.59 (Rahman and Khalequzzaman, 2004). Sexual asynchrony occurs in the eclosion of male and female moths. Males emerge early in the scotophase, followed by females in the later part of scotophase. Adults are generally short lived and do not seem to disperse far from the emergence site (Harris, 1989).

Chilo auricilius lays eggs in clusters on the lower surface of leaves. The first and second instars feed on leaves. Later larval instars bore inside the stalks causing deadheart. It is primarily a pest of sugarcane and rice. There are usually three or four distinct generations observed with overwintering larvae.

Chilo infuscatellus has five generations a year with a larval diapause during winter in northern India, and in southern India, it passes six generations a year without any overwintering phase (Harris, 1990). Females oviposit on the underside of the leaf surface. Eggs hatch in 5–9 days. Early instars feed on leaves and usually third instars enter the stems.

Chilo sacchariphagus lays eggs on the undersurface of the leaf. First-instar larvae feed gregariously in the leaf whorl. Second instars bore into the stalk and make galleries within the internodal regions.

Ostrinia furnacalis lays eggs in several masses on the undersurface of the leaf near the midrib. The first- and second-instar larvae feed on young leaves. Third-instar larvae bore into the stalk through the midribs and sheaths and continue to feed on stalks until pupation. The egg, larval, pupal, and adult stages last for 3–4, 17–20, 5–6, and 4–7 days, respectively.

Coniesta ignefusalis is a serious pest on pearl millet in the Sahel region. The female lays eggs between the leaf sheath and the stem in groups of about 20–50 with a total egg-laying capacity of over 200. Two or three generations occur in a crop-growing season. Larvae can migrate between plants (Youm et al., 1996). During the dry season, the final-instar larva loses all integumental pigment and enters into facultative diapause (Harris, 1962). The adult moths are not known to have a migratory habit.

Maliarpha separatella lays eggs in groups on the upper surface of the leaf and the eggs are cemented to the leaf surface by a sticky substance (Heinrichs and Barrion, 2004). After hatching, the larvae feed on green tissues in the leaf sheath for few days before boring into the stalk and feeding on the internodal contents. The larval period lasts for about 45–60 days. The last-instar larva cuts out an exit hole for the adult and pupates inside the stalk. The pupa lasts for over 12 days and the whole life cycle lasts for 40–90 days (Breniere et al., 1962). Facultative larval diapause is known to occur in *M. separatella* during dry seasons.

Eldana saccharina is primarily a sugarcane borer and its female lays eggs in batches of 50–100 in the folds of dead leaves, in leaf sheaths, or on dry leaves. The egg period lasts for about 4–6 days. First-instar larvae feed on leaves and second instars bore into the plant stalk and feed in the internodal area by making tunnels within which they pupate. Diapause is not known to occur in *E. saccharina*.

S. inferens lays a total of around 300 eggs in batches and, unlike *Chilo*, the eggs are placed safely between the leaf sheath and the stem (Figs. 3.10 and 3.11). A single egg batch has 75–100 eggs. Oviposition occurs over a period of weeks. The egg period lasts for 6–10 days. In a batch of eggs, one or few of the newly hatched larvae can bore into the stem without coming to the surface of the plant and the remaining larvae disperse to neighboring plants (Fig. 3.12). A single larva can infest more

FIGURE 3.10

Egg mass of *Sesamia inferens* laid within the leaf sheath of finger millet.

Courtesy of A. Kalaisekar.

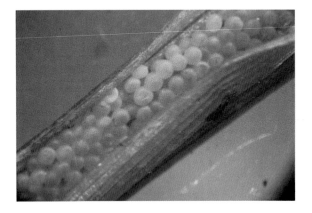

FIGURE 3.11

Egg mass of *Sesamia inferens*.

Courtesy of A. Kalaisekar.

FIGURE 3.12

First-instar larvae of *Sesamia inferens* within the leaf sheath.

Courtesy of A. Kalaisekar.

than one stem. The frass materials produced during larval feeding can be seen through openings in the leaf sheath. The total larval period ranges from 25 to 60 days (Rothschild, 1971). The larva pupates within the stem or in the withered leaf sheaths. The final instar enters into diapause under dry conditions or cold temperatures. The pupal period lasts about 10–15 days. Adult moths exhibit strong flight abilities and can fly up to 50 km. There are three and six generations in temperate and tropical climates, respectively.

S. calamistis larvae are cannibalistic, which causes high rates of mortality (Bosque-Perez and Dabrowski, 1989; Sallam et al., 1999).

B. fusca is the most important borer pest, other than *C. partellus*, in grain sorghum in Africa (Kfir et al., 2002). *B. fusca* lays eggs in batches of about 20–100 between the leaf sheath and the stem. The egg period lasts for about 1 week. After hatching, the larvae initially disperse over plants before they enter the leaf whorls and start feeding on the leaves. Later-instar larvae bore into the stem and feed for 20–35 days, producing extensive tunnels in internodal areas. During dry or cold seasons the larvae enter into a diapause of 6 months or more in stems, stubble, and other plant residues. They pupate in the tunnels, after excavating an emergence hole to facilitate the exit of the adult moth. The pupal period takes about 9–14 days. The total life cycle is completed in 7–8 weeks when conditions are favorable (Harris and Nwanze, 1992).

Mythimna separata lays about 800 eggs after a preovipositioning period of 2–5 days. The egg period lasts for about 4–5 days. There are four larval instars and the total larval period lasts for 12–15 days. Pupal periods last for 8–12 days. The total life cycle takes about 30–40 days.

SHOOT BUG

The *P. maidis* female inserts eggs into young tissues around the midrib of the leaf. Each female lays 500–600 eggs. The egg period lasts for about 7–9 days. There are five nymphal instars with each lasting for about 8–11 days. The adults live for about 45–75 days depending upon the weather and the host conditions (Rioja et al., 2006). However, the duration of various stages fluctuates highly under different temperature regimes (Tsai and Wilson, 1986) and according to the host (Singh and Seetharama, 2008).

The preovipositioning and ovipositioning periods range from 1 to 3 days and 6 to 7 days on sorghum, respectively. The total fecundity of brachypterous and macropterous forms ranges from 18 to 94 and 5 to 64, respectively, on sorghum. The total nymphal period averages 22–26 days on sorghum (Chelliah and Basheer, 1965; Rajasekhar, 1989; Singh and Seetharama, 2008). The total life cycle is completed in about 22 days on corn (Catindig et al., 1996) under optimum weather conditions. There are fitness advantages existing between the two dimorphic wing forms of *P. maidis*. Brachypterous females are more abundant and reproductively more efficient in the offspring developing faster, shorter preovipositioning period, higher number of eggs per female, longer ovipositioning period, and greater longevity (Fernández-Badillo, 1988; Singh and Seetharama, 2008).

Nymphs and adults colonize in the leaf whorls and sheaths. The infestation may result in discoloration, reddening, twisting of the topmost leaves, and withering and drying of leaves, and may prevent emergence and development of the panicle (Singh and Rana, 1992; Singh, 1997). Under severe infestations, the entire plant dries up.

HEAD BUG

Calocoris angustatus inserts 150–200 cigar-shaped eggs under the glumes or in between the anthers of florets. The egg period lasts for about 5–7 days and the nymphs feed on milky grains. There are five nymphal instars taking about 20 days to become an adult (Ghose, 2008). There are some predaceous bugs, viz., the lygaeid bug *Geocoris tricolor* and the raduvid bug *Reduviolus* sp., found in large numbers amid the head bugs.

Nymphs and adults of *C. angustatus* feed on the developing grains, which become tanned and shriveled. Under severe infestation, the panicles become blasted and dried up with no grain. Head bug damage increases the severity of grain molds and deteriorates the grain quality (Steck et al., 1989; Sharma et al., 2000). As a result, the grains produced during the rainy season are the worst affected by head bugs and grain molds, and as a result such grains gain a lower market price than those produced during the postrainy season (Duale and Nwanze, 1999).

The mirid bugs *Eurystylus bellevoyei* and *Creontiades pallidus* infest sorghum earheads, in addition to *C. angustus*, in India and Africa. The life cycle is completed in 14–16 and 17–23 days by *E. bellevoyei* and *C. pallidus*, respectively, and the bugs are active during September–October (Sharma and Lopez, 1990a,b).

NEZARA VIRIDULA

The female lays about 300 barrel-shaped eggs in clusters on both surfaces of leaves. The eggs are whitish or yellowish and turn pinkish when nearing hatching. The egg period takes about 5–7 days. Nymphs are brownish red in color with multicolor spots. The nymphs have five instars. The total nymphal period takes about a month. There are four or five overlapping generations in a year.

Nymphs and adults suck the cell sap from all tender parts of the plant and developing grains.

CICADULINA MBILA

Females lay their creamy-white eggs in the leaf tissue, usually near the midrib. The eggs take a week to over a month to hatch. There are five nymphal instars. The nymphal period takes about 2–3 weeks. The preovipositioning period ranges from 2 to 20 days. Fecundity is around 110–130. Adult females developed on sorghum live for about 20 days and on pearl millet for about 30 days. The life cycle of *C. mbila* is highly influenced by weather factors, host plant, and population origin (Rose, 1973a,b,c; Okoth et al., 1987; Bosque-Perez and Alam, 1992).

C. mbila is known to transmit maize streak virus (MSV), a geminivirus in maize and sorghum.

Nymphs and adults suck the sap from leaves. Direct feeding damage by this leafhopper causes insignificant damage, but their ability to transmit MSV is a cause of great concern especially at early to middle plant growth stages. *C. mbila* can acquire MSV from infested plants within 15 s of feeding (Storey, 1938). The symptoms of MSV appear in the newly formed leaves, initially as spherical chlorotic spots, and such spots enlarge to coalesce into longitudinal chlorotic streaks (Bosque-Perez et al., 1998).

BLISSUS LEUCOPTERUS

Chinch (a Spanish word meaning "pest") bugs have caused great crop losses to sorghum and corn in the United States for several decades (Fitch, 1855; Osborn, 1888; Hayes, 1922; Dahms et al., 1936; Dahms,

1948; Kerr, 1966; Leonard, 1966; Busey and Zaenker, 1992; Heng-Moss et al., 2003; Yue et al., 2000; Anderson et al., 2006; Ramm et al., 2015). The chinch bug genus comprises grass-feeding bugs. In the United States, there are four species of chinch bugs that cause economic losses to the crops: chinch bug, *B. leucopterus leucopterus*; southern chinch bug, *Blissus insularis*; hairy chinch bug, *B. leucopterus hirtus*; and western chinch bug, *Blissus occiduus* (Vittum et al., 1999; Anderson et al., 2006).

Usually chinch bugs undergo two generations in a year, and in temperate regions, adults overwinter. The nymphs must molt into adult bugs before the start of winter for overwintering. The nymphs that fail to transform into adults before the start of winter cannot survive (Andre, 1937; Leonard, 1966). Mating takes place after breaking from the hibernation. The preovipositioning period in *B. insularis* is 7–10 days. A single female lays about 500–600 eggs. Egg hatches in about 2 weeks' time. Developmental time varies with locality. A generalized generation cycle for 1 year is given in Fig. 3.13.

There are five nymphal stages observed in *B. insularis*. Adult longevity for males and females is 42 and 70 days, respectively (Kerr, 1966).

The leaf-footed coreid bug *Leptoglossus phyllopus* is a potential carrier of sorghum fungal pathogens affecting developing grains (Louis and Ramasamy, 2008).

APHIDS

Life cycles of aphids are extremely complicated and difficult to understand. In general, aphid life cycles are complex and highly influenced by the climate, the host plant availability, and its own population pressure. The life stages are loaded with several morphological forms. Aphids are extremely adaptable to the climatic conditions and hosts in the locations where they live. There are two types of life cycle in

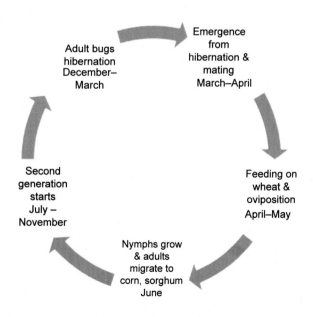

FIGURE 3.13

Annual generation cycle of chinch bug in the United States.

Courtesy of A. Kalaisekar.

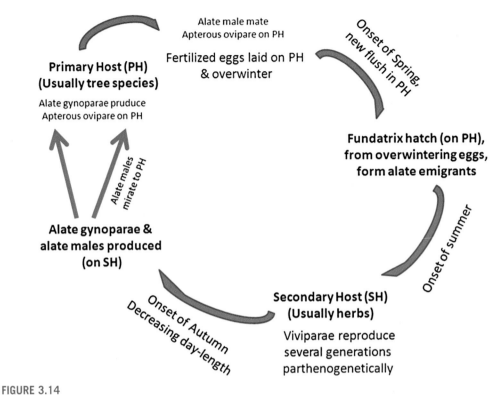

FIGURE 3.14

Holocyclic life cycle of aphids.

Courtesy of A. Kalaisekar.

aphids, viz., holocyclic and anholocyclic. A typical holocyclic life cycle involves mating between sexual adults and oviparous females laying fertilized eggs on the primary host, trees or shrubs. Such eggs overwinter and hatch after the winter is over. Individuals thus produced migrate to secondary hosts, which are usually herbs and unrelated to the primary host plant. On the secondary host plant, aphids pass several generations parthenogenetically by directly producing young ones and no eggs are laid. These asexually produced populations move to the primary host as winter approaches and produce sexually active male and females. Such a cyclical life cycle involving sexual and parthenogenetic lifestyles is called a holocyclic life cycle (Fig. 3.14). The anholocyclic life cycle is totally devoid of a sexual life phase and the generations pass through by means of parthenogenetic viviparous individuals (Fig. 3.15). The aphid species living in temperate regions, especially the Northern Hemisphere, invariably undergo a holocyclic life cycle, whereas the tropical-living aphids undergo anholocycly. Owing to genetic recombination through sexual reproduction, the species diversity of aphids is very high in the temperate region.

Most species of aphids are adapted to an arboreal lifestyle and a small fraction experiences the herbaceous life. In general, tree-dwelling aphids undergo a holocyclic life cycle almost entirely in temperate regions of the world, whereas in tropical regions, aphids lose the sexual phase and are almost

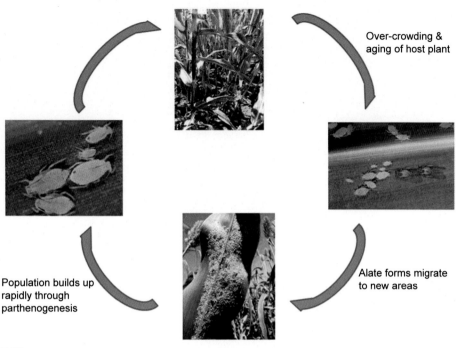

FIGURE 3.15

Anholocyclic life cycle of aphids (e.g., *Melanaphis sacchari* on sorghum in Hyderabad, India).

Courtesy of A. Kalaisekar.

entirely anholocyclic with a characteristically shortened life cycle through parthenogenetic viviparous adults. Therefore, anholocycly is used for species that cannot reproduce as eggs in the absence of a suitable alternate host plant or through other environmental circumstances (Mordvilko, 1935; Hille Ris Lambers, 1966).

Aphids that survive on a single species or a group of related species of plant host are called monoecious. Heteroecious aphids survive on two different groups of host plant species at different times of the year. Almost all the millets are cultivated in tropical regions of the world, except the proso millet, which is cultivated in temperate regions as well. Therefore, the aphids that attack millets are anholocyclic by virtue of the growing environment of the host plants. Most of the information on aphids in this section are of Blackman and Eastop, 1984a,b, 1990, 1994a,b, 2000a,b, 2006, 2007a,b. A list of aphids that attack various millet crops is given in Table 3.1.

Melanaphis sacchari

It is mainly or entirely anholocyclic in most of its areas of occurrence. There is another closely resembling species, *Melanaphis sorghi* (Theobaldi), which usually feeds on *Sorghum halepense* (Blackman and Eastop, 1990). *M. sacchari* adults are either apterae or alate. The viviparous adults produce young ones parthenogenetically (Varma et al., 1978; Singh et al., 2004).

Table 3.1 Aphid Species Recorded on Various Millets (Blackman and Eastop, 1994a,b, 2000a, 2006, 2007a)

Millet species	Aphid Species
Sorghum bicolor	1. *Anoecia corni* 2. *Anoecia fulviabdominalis* 3. *Anoecia krizusi* 4. *Aphis fabae* 5. *Aphis gossypii* 6. *Diuraphis noxia* 7. *Forda hirsuta* 8. *Forda orientalis* 9. *Geoica lucifuga* 10. *Hysteroneura setariae* 11. *Melanaphis sacchari* 12. *Melanaphis sorghi* 13. *Myzus persicae* 14. *Paracletus cimiciformis* 15. *Rhopalosiphum maidis* 16. *Rhopalosiphum padi* 17. *Schizaphis graminum* 18. *Sipha flava* 19. *Sipha maydis* 20. *Sitobion africanum* 21. *Sitobion avenae* 22. *Sitobion leelamaniae* 23. *Sitobion miscanthi* 24. *Tetraneura capitata* 25. *Tetraneura chui* 26. *Tetraneura fusiformis* 27. *Tetraneura javensis* 28. *Tetraneura nigriabdominalis* 29. *T. nigriabdominalis* subsp. *shanxiensis* 30. *Tetraneura triangula* 31. *Tetraneura ulmi*
Setaria italica	1. *A. corni* 2. *M. sacchari* 3. *P. cimiciformis* 4. *R. maidis* 5. *R. padi* 6. *Rhopalosiphum rufiabdominale* 7. *Schizaphis graminum* 8. *Sipha elegans* 9. *S. flava* 10. *S. avenae* 11. *S. miscanthi* 12. *Tetraneura caerulescens* 13. *T. fusiformis* 14. *Tetraneura javensi* 15. *T. nigriabdominalis* 16. *Tetraneura yezoensis*

Table 3.1 Aphid Species Recorded on Various Millets—cont'd

Millet species	Aphid Species
Echinochloa frumentacea	1. *A. corni* 2. *A. fulviabdominalis* 3. *R. maidis* 4. *R. padi* 5. *R. rufiabdominale* 6. *S. flava* 7. *T. fusiformis* 8. *T. nigriabdominalis* 9. *T. yezoensis*
Eleusine coracana	1. *A. corni* 2. *A. gossypii* 3. *Aphis spiraecola* 4. *Brachycaudus helichrysi* 5. *G. lucifuga* 6. *H. setariae* 7. *M. sorghi* 8. *R. maidis* 9. *R. padi* 10. *S. graminum* 11. *S. maydis* 12. *S. avenae* 13. *S. leelamaniae* 14. *S. miscanthi* 15. *Tetraneura basui* 16. *T. fusiformis* 17. *T. javensis* 18. *T. yezoensis*
Pennisetum glaucum	1. *A. corni* 2. *Anoecia cornicola* 3. *A. fulviabdominalis* 4. *Anoecia vagans* 5. *A. gossypii* 6. *F. hirsuta* 7. *F. orientalis* 8. *Geoica utricularia* 9. *H. setariae* 10. *M. sacchari* 11. *Protaphis middletonii* 12. *R. maidis* 13. *R. padi* 14. *R. rufiabdominale* 15. *S. graminum* 16. *S. elegans* 17. *S. maydis* 18. *S. avenae* 19. *S. leelamaniae* 20. *S. miscanthi* 21. *Sitobion pauliani* 22. *Tetraneura africana* 23. *T. basui* 24. *T. fusiformis* 25. *T. yezoensis*

Continued

Table 3.1 Aphid Species Recorded on Various Millets—cont'd

Millet species	Aphid Species
Panicum miliaceum	**1.** *A. corni* **2.** *A. vagans* **3.** *Forda marginata* **4.** *G. utricularia* **5.** *Melanaphis pyraria* **6.** *M. sacchari* **7.** *M. persicae* **8.** *Neomyzus circumflexus* **9.** *P. cimiciformis* **10.** *P. cimiciformis* subsp. *panicumi* **11.** *R. maidis* **12.** *Rhopalosiphum nymphaeae* **13.** *R. padi* **14.** *S. graminum* **15.** *S. maydis* **16.** *S. avenae* **17.** *T. caerulescens* **18.** *T. javensis* **19.** *T. fusiformis* **20.** *T. ulmi* **21.** *T. yezoensis*
Panicum sumatrense	**1.** *H. setariae*
Paspalum scrobiculatum	**1.** *A. gossypii* **2.** *G. lucifuga* **3.** *H. setariae* **4.** *R. maidis* **5.** *Sitobion graminis* **6.** *Sitobion indicum* **7.** *T. basui* **8.** *T. fusiformis*
Eragrostis tef	**1.** *H. setariae* **2.** *S. graminum* **3.** *Smynthurodes betae* **4.** *T. ulmi*

It has four nymphal instars and takes about 4–13 days to mature into an adult (Chang, 1981). Adults are known to survive for 10–30 days. A single female aphid produces around 50–80 young in its life-time. Alate females produce fewer progeny than the apterae (van Rensburg, 1973; Meksongsee and Chawanapong, 1985). The total life cycle takes less than a week in the dry season. *M. sacchari* gener-ally prefers older leaves. Under severe infestations the entire plant looks withered with dried-up leaves and copious amounts of honeydew secreted by aphids are found on the leaves and on the ground. The population rapidly increases during the dry spell after the rains.

M. sacchari is not entirely anholocycly and there is evidence (Yadava, 1966; Setokuchi, 1975; David and Sandhu, 1976) to suggest that this aphid produces oviparous morphs called oviparae also. In

M. sacchari, oviparae are normally seen after the 16th generation during November–December and February–March, especially between 30 and 41° latitudes. Oviparae are found on sorghum and on Amur silver-grass (*Miscanthus sacchariflorus*) when the average temperatures are around 18–20°C.

Rhopalosiphum maidis

It is an entirely anholocyclic aphid species and its distribution is largely restricted to tropical and warm temperate regions of the world. There are some reports of the existence of heteroecious holocycly in Pakistan, Japan (Torikura, 1991), and Korea (Lee et al., 2002). *R. maidis* is Asiatic in origin and is not known to survive in temperate climatic regions. It is a most important aphid pest of cereals in tropical and warm temperate climates (Blackman and Eastop, 2000a,b). Apterae are elongated and yellow-green or dark olive green or bluish green. Alate females fly from off-season hosts to find sorghum or maize. After settling on a suitable host, they give birth to young ones parthenogenetically, which pass through four nymphal stages. The nymphal period lasts for about 6–8 days. Apterous parthenogenetic females live for about 20–30 days. During winter, the aphids migrate to winter cereals and wild grasses for wintering (Foott, 1977; Sumner et al., 1986). When the colony becomes stressed or overcrowded winged forms are produced. Males are rare, and wingless forms are the most common ones. There are 40–50 generations produced in a year.

Anoecia corni

It has a holocyclic life cycle in temperate regions of Europe, central and eastern Asia, Africa, Argentina (Ortego, 1998), and North America (Blackman and Eastop, 2000a,b). The primary hosts are dogwood, *Cornus* spp., and the secondary hosts are grasses of Poaceae and sedges of Cyperacae. The oviparae (egg-laying sexual females) lay their eggs on the bark of the trunk of *Cornus* spp. On shifting to secondary hosts, the aphids colonize the roots of grasses or sedges including cultivated cereals and millets. Similar to the majority of aphids, *A. corni* also undergoes an anholocyclic life cycle in the tropics.

Anoecia cornicola

This follows holocycly on its primary hosts *Cornus amomum*, *Cornus stolonifera*, and the roots of Poaceae in North America. It is anholocyclic in Brazil on grass roots (Blackman and Eastop, 2000a,b).

Anoecia fulviabdominalis

This aphid is holocyclic in temperate regions. Its primary hosts are *Cornus* spp. and secondary hosts are grasses (roots of Poaceae). It is anholocyclic in warm climates and lives on roots of grasses. *A. fulviabdominalis* is a major pest of upland rice in east and southeast Asia. It is also recorded on roots of many cultivated millets.

Anoecia krizusi

This is monoecious holocyclic on grass roots. It is an important root pest on sorghum and is distributed in Europe and Asia.

Anoecia vagans

It is holocyclic, its primary hosts are *Cornus* spp., and its secondary hosts are roots of Poaceae. It is a pest of many cereals and millets in Europe, east Siberia, west Asia, and India.

Aphis fabae

This species is holocyclic in temperate Europe, and its primary host is the guelder-rose, *Viburnum opulus* (a perennial deciduous shrub). Secondary hosts are several fabaceous plants. There are primarily four species complexes recognized based on their secondary hosts. It is anholocyclic on fabaceous plants and on millets in southern Europe, Africa, southwest Asia, south and east Asia, the Indian subcontinent (Kim et al., 2006), South America, Hawaii, and the Auckland Isles (Blackman and Eastop, 2007a,b).

Aphis gossypii

This is a common aphid species in the tropics. During holocycly, unrelated plants are used as primary hosts in Asia. But in China, monoecious holocyclic populations are observed on malvaceous plants, particularly, cotton and *Hibiscus*. It is anholocyclic in warm climates on malvaceous plants and many millet crops such as finger millet, pearl millet, sorghum, and kodo millet.

Aphis spiraecola

Usually a tropical anholocyclic species, *A. spiraecola* is holocyclic on rosaceous plants, *Spiraea* spp., as primary hosts in North America, Brazil, and Japan (Blackman and Eastop, 2007a,b). It is reported on several unrelated plant species across over 20 families and is a major pest of *Citrus* (Blackman and Eastop, 2000a,b) and commercial yarrow, *Achillea millefolium* (=*collina*) (Gama et al., 2010). It is recorded on finger millet.

Brachycaudus helichrysi

It is heteroecious holocyclic. Its primary hosts are *Prunus* spp. and its secondary hosts are several unrelated plant species including Compositae. It is recorded on finger millet.

Ceratovacuna lanigera

It is a tropical species and is anholocyclic throughout its area of occurrence. *C. lanigera* forms dense colonies on the underside of leaves of several species of Poaceae. *C. lanigera* is a serious pest of sugarcane in southeast Asia and India (Joshi and Viraktamath, 2004).

Diuraphis noxia

It is monoecious holocyclic in temperate Asia. It is entirely anholocyclic throughout its area of occurrence in North America (Brewer and Elliot, 2004) and in warmer regions. It survives on cereals, millets, and grasses. It is a Palearctic species and now a widespread one.

Forda formicaria

The primary hosts are *Pistacia* spp. and the secondary hosts are grasses. On *Pistacia* spp., the aphid forms half-moon-shaped leaf galls. On grasses and sedges, *F. formicaria* feeds on the roots, where the tending ants visit the aphid colony. Unusually, *F. formicaria* often overwinters in the nests of tending ants during the anholocyclic phase of its life cycle. This is why the species was named *formicaria* in 1837.

Forda hirsuta

It is heteroecious holocyclic in temperate regions, like other *Forda* species. Its primary host *Pistacia* spp. It is anholocyclic on the roots of several poaceous grasses including cultivated cereals and millets.

Forda marginata

It is heteroecious holocyclic in southern Europe, the Middle East, and northwest India. It is anholocyclic on grass roots and overwinters in the nests of ants (Blackman, 1987; Chakrabarti and Raychaudhuri, 1978).

Forda orientalis

It is anholocyclic on roots of grasses. Its sexual phase is not known.

Geoica lucifuga

It is anholocyclic on roots of grasses and sedges. Its sexual phase is not known.

Geoica utricularia

The biology of this species is quite similar to that of *Forda* spp. It is heteroecious holocyclic. Its primary host is *Pistacia* spp., forming leaf galls, and its secondary hosts are grasses (on roots).

Hyalopterus pruni

Its alatae have a green abdomen and white wax patches segmentally. Apterae are elongated with darker green mottling. It is cosmopolitan and occurs on *Prunus* species. It occasionally feeds on cereals and millets.

Hysteroneura setariae

It is heteroecious holocyclic in temperate North America. Its primary host is *Prunus domestica*. Secondary hosts are grasses, cereals, and millets (Blackman and Eastop, 2000a,b). It is anholocyclic on grasses. *H. setariae* colonizes near the basal parts of panicles of several species of Poaceae.

Melanaphis pyraria

It is heteroecious holocyclic in temperate regions. Its primary host is *Pyrus communis* and secondary hosts are plants of Poaceae.

Melanaphis sorghi

This is found living on *Sorghum bicolor* and also on other grasses. It is almost entirely anholocyclic throughout its areas of occurrence with an exception of a report of holocycly on sorghum in Japan (Setokuchi, 1975)

Metopolophium dirhodum

Apterae are green or yellowish green with a brighter green spinal stripe. Each segment of antennae has dark apices (Blackman and Eastop, 2000a,b). *M. dirhodum* is heteroecious holocyclic across temperate Europe on *Rosa* species. Alates migrate to several species of Poaceae and Cyperaceae during warmer seasons and reproduce parthenogenetically. *M. dirhodum*, in contrast to *R. maidis*, is a major pest of cereals and has become widely distributed in temperate regions of the world (Blackman and Eastop, 2000a,b). The populations of *M. dirhodum* found in the warmer parts of the world are entirely anholocyclic.

Myzus persicae

It is heteroecious holocyclic. Its primary host is *Prunus persica* and secondary hosts include various plant species of over 40 plant families (Blackman and Eastop, 2000a,b). It is anholocyclic on several unrelated plant species.

Neomyzus circumflexus

It is entirely anholocyclic with no sexual morphs recorded so far. Extremely polyphagous with a host range cutting across the plants from monocots to dicots, gymnosperms, and pteridophytes.

Paracletus cimiciformis

It is heteroecious holocyclic. Its primary hosts are *Pistacia* spp. (forms leaf galls) and secondary hosts are grasses (living on the roots). It is anholocyclic on roots of grasses.

Protaphis middletonii

It is heteroecious holocyclic. It lives on roots of species of several plant families including Apiaceae, Brassicaceae, Compositae, Lamiaceae, and Poaceae.

Rhopalosiphum nymphaeae

It is heteroecious holocyclic. Its primary hosts are *Prunus* spp. and secondary hosts consist of several species of water plants. There are records of *R. nymphaeae* having the ability to survive underwater. *R. nymphaeae* was employed as a biocontrol agent against water weeds in rice (Oraze and Grigarick, 1992).

Rhopalosiphum padi

It is heteroecious holocyclic. Its primary hosts are *Prunus padus* in Europe and *P. padus*, *Prunus grayana*, and *Prunus ssiori* in Japan (Torikura, 1991). Secondary hosts include the plants of Cyperaceae, *Iris*, Juncaceae, and Typhaceae.

Rhopalosiphum rufiabdominale

It is heteroecious holocyclic. Primary hosts are rosaceous plants, viz., *Prunus* spp. and *Rhodotypos scandens*, and the secondary hosts are species of Poaceae and Cyperaceae and cultivated solanaceous plants such as potato, tomato, and capsicums (forms colonies on underground parts). It is anholocyclic in warmer climates.

Schizaphis graminum

S. graminum is popularly known as the "greenbug" and has been a major pest of small grains for over 150 years in North America (Nuessly and Nagata, 2005). Its ability to develop biotypes has been a constant threat to the development of resistant cultivars of sorghum and other small-grained cereals and millets in the United States (Michels, 1986; Nagaraj et al., 2002).

Similar to most of the aphids, the greenbug also undergoes holocycly in temperate North America and anholocycly in warm-climate regions. However, in contrast to several other species of holocyclic aphids, *S. graminum* is monoecious as it exclusively develops on true grasses (Poaceae) and does not alternate to a nongrass host during holocycly (Webster and Phillips, 1912; Blackman and Eastop, 2007a,b; Royer et al., 2015). Both modes of life cycles are exclusively associated with grasses and there is no host alteration to nongrass plants. The "sorghum-adapted forms" of *S. graminum* produce sexually in warm latitudes and, therefore, most probably the greenbug could have originated from farther south in Europe or Asia (Blackman and Eastop, 2007a,b). During holocycly, after mating the females lay eggs predominantly on *Poa pratensis* and the eggs overwinter (Nuessly and Nagata, 2005; Blackman and Eastop, 2007a,b). Overwintering is observed in nymphs and adults also (Royer et al., 2015). In anholocycly, adult greenbugs reproduce parthenogenetically on grasses. A biotype (Florida isolate) of the greenbug was found to produce one to five nymphs per day on seashore *Paspalum* (Nuessly and Nagata, 2005).

The biology of *S. graminum* is highly influenced by temperature. The greenbugs supercool at −26°C and, therefore, survival below this temperature is not possible for *S. graminum* (Jones et al., 2008).

S. graminum is known to exist in several biotypic forms (Harvey and Hackerott, 1969; Weng et al., 2010). Biotypes have been defined by their ability to damage different plant genotypes. There are 22 greenbug biotypes so far identified and the biotypes C, E, and I cause significant economic losses in cultivated cereals and millets, especially wheat and sorghum (Puterka and Peters, 1990; Shufran et al., 1997, 2000; Burd and Porter, 2006; Weiland et al., 2008).

Greenbugs are vectors of viral disease, viz., sugarcane mosaic (Ingram and Summers, 1938), barley yellow dwarf (Murphy, 1959), and maize dwarf mosaic (Nault and Bradley, 1969).

Sipha elegans

It is monoecious holocyclic on upper sides of leaves of various grasses, millets, and cereals. It is distributed in temperate regions of the world.

Sipha flava

It is monoecious holocyclic on leaves of grasses and sedges, and anholocyclic in warmer climates. The males are apterous in areas with cold winters and anholocyclic in warmer climates (Blackman and Eastop, 2000a,b). Nymphs and adults suck the sap from leaves of several grasses including cultivated cereals and millets such as *Hordeum*, *Triticum*, *Panicum*, *Paspalum*, *Pennisetum*, *Setaria*, and *Sorghum*. Colonies of *S. flava* are not attended by ants.

Nymphs undergo four instars. Development from nymph to reproducing adult takes about 8 days on *S. bicolor*. More often an *M. sacchari* colony is noticed amid *S. flava* on sugarcane leaves. Interestingly, *S. flava* does not produce an alarm pheromone, but it responds to the alarm pheromone of *M. sacchari* by quickly falling from the plant (Nuessly, 2011).

Sipha maydis

It is monoecious holocyclic on Poaceae. It forms colonies on the upper side of leaves, stems, or inflorescences. It is anholocyclic in warmer climates.

Sitobion africanum

It is mainly anholocyclic. It survives mainly on plants of Poaceae in Africa, Indian Ocean islands, and Yemen.

Sitobion avenae

It is monoecious holocyclic on various species of Poaceae and occasionally on some dicots. This aphid prefers to feed on the upper leaves and on the ears. Anholocyclic overwintering is common.

Sitobion graminis

It is mostly anholocyclic on grasses and sedges.

Sitobion indicum

It is entirely anholocyclic on orchids in India.

Sitobion leelamaniae

It is mostly anholocyclic on many species of grasses and cereals in southern India, southern Africa, and Sri Lanka. It is heteroecious holocyclic in cold regions with *Hagenia abyssinica* as primary host.

Sitobion miscanthi

It is almost entirely anholocyclic on Poaceae and *Cyperus*, and sometimes found on semiaquatic species, *Polygonum hydropiper* (Eastop, 1966). There are some reports of the presence of oviparae and males in India (David, 1976) and Australia (Hales et al., 2010). Apterae are green or reddish brown or dark brown. It largely feeds on species of Poaceae.

Sitobion pauliani

It is almost entirely anholocyclic on inflorescences of grasses, especially *Pennisetum*, *Setaria*, and *Eleusine* in Africa, Madagascar, the Andaman and Niccobar Isles (India), Sri Lanka, southeast Asia, and Central and South America.

Smynthurodes betaesssss

It is heteroecious holocyclic on *Pistacia* spp. It is host-alternating, with a 2-year cycle. Secondary hosts are particularly Compositae, Leguminosae, and Solanaceae, and rarely Poaceae and Cyperaceae. It is anholocyclic in warm areas.

Tetraneura nigriabdominalis

It is a holocyclic aphid in temperate regions. The primary host is elm, *Ulmus* species. It effects host alternation between leaf galls on *Ulmus* and roots of Poaceae. The leaf galls made by *T. nigriabdominalis* on elm are stalked, hairy, elongate, pouchlike outgrowths on the upper side of leaves. Alates leave the elm gall through lateral slits in May–July and found colonies on the roots of plants of Poaceae such as *Cynodon*, *Digitaria*, *Echinochloa*, *Oryza*, *Saccharum*, and *Setaria*. Apterae on grass roots are greenish or brownish white. The colonies formed on roots are attended by ants. This species is known particularly as a pest of rice (Blackman and Eastop, 2000a,b) and found feeding on the roots of upland rice plants in west Africa (Akibo-Betts and Raymundo, 1978). In Japan, the species is heteroecious holocyclic between *Ulmus* and the roots of Gramineae.

T. nigriabdominaalis is anholocyclic in India and found throughout the year on various cultivated and wild grasses. It is a serious pest on finger millet roots in south India. In May–September it occurs mainly on *Eleusine coracana*, June–September on sorghum, and during cropping off-seasons, wild grass, *Eragrostis tenella*, offers shelter for *T. nigriabdominalis*. When adults and nymphs feed persistently on *E. coracana*, the entire collar region of the root turns back, the cortical tissues dry up, and the secondary roots show a burnt appearance. It passes through four nymphal instars, the nymphal period lasting 7–9 days. Adult females live for 5–11 days and produce 10–35 offspring each (Gadiyappanavar and Channabasavanna, 1973). The aphid produces honeydew that attracts ants. The tending ants transport the aphids from plant to plant and protect the colonies from predators and parasites (Gadiyappanavar and Channabasavanna, 1973; Wijerathna and Edirisinghe, 1995).

Tetraneura triangula

It is anholocyclic on roots of various plant species of Poaceae.

Tetraneura africana
It is anholocyclic on roots of grasses and sedges (especially on *Cynodon* roots). It is found on roots of sorghum.

Tetraneura basui
It is anholocyclic on roots of Poaceae. It is found in northeast India (Raychaudhuri, 1980).

Tetraneura caerulescens
It is holocyclic. Its primary hosts are *Ulmus* spp. and secondary hosts are roots of Poaceae.

Tetraneura capitata
It is found on roots of sorghum.

Tetraneura longisetosa
It is holocyclic. Its primary hosts are *Ulmus* spp. in Europe. It is anholocyclic on roots of sorghum in Asia.

Tetraneura fusiformis
It is heteroecious holocyclic in Japan. Its primary host is *Ulmus japonica*. Secondary hosts are roots of grasses. It is anholocyclic in tropical Asia, Africa, and South and Central America.

Tetraneura javensis
It is heteroecious holocyclic in temperate Asia and anholocyclic in tropical Asia on the roots of Poaceae. It is most common on roots of sugarcane.

Tetraneura ulmi
It is heteroecious holocyclic in Europe, North America, and temperate Asia. It is host alternating between *Ulmus* spp. and roots of Poaceae. It is anholocyclic in tropical Asia on the roots of several species of grasses including cultivated cereals and millets.

Tetraneura yezoensis
It is heteroecious holocyclic in temperate Asia and anholocyclic in tropical Asia and Australia on the roots of Poaceae.

MIDGES

The sorghum grain midge, *S. sorghicola*, is a serious pest throughout the areas of sorghum cultivation in the world. As early as the 1890s the midge was so serious that it wiped out the seed sorghums in many southern states of the United States (Hansen, 1923; Walter, 1941). Around the same period in 1898, Coquillett described the sorghum midge as *Diplosis sorghicola*.

Gall midges are widespread pests of several cultivated cereals and millets apart from wild grasses (Walter, 1941; Harris, 1976; Harris et al., 2003; Ahee et al., 2013). Several species of gall midges belonging to the subfamily Cecidomyiinae form galls in plants, and the sorghum midge, *S. sorghicola*, does not form galls in association with florets (Ahee et al., 2013). *S. sorghicola* substantially reduces global sorghum yields (Sharma, 1993; Sharma et al., 1999; Damte et al., 2009).

Males emerge first from the pupae and hover around the panicles waiting for the female to emerge. Emergence of both the sexes takes place in the morning hours. Females are reported to release sex pheromones to attract males (Sharma and Vidyasagar, 1992). After mating, females search for sorghum panicles at anthesis for ovipositioning (Sharma, 2004). Females prefer the panicles at midanthesis for ovipositioning (Modini et al., 1987). A single female lays about 50–100 eggs, generally one egg in a spikelet and occasionally about three to five eggs per spikelet (Palma, 1988; Sharma, 2004). Although not more than one egg is deposited at a time in a spikelet, it is common for several females to oviposit within the same spikelet (Gable et al., 1928). Ovipositioning is completed by midday and most of the female flies die before sunset (Sharma, 2004). However, there are reports that the females of *S. sorghicola* exhibit diurnal oviposition activity and preference for panicles at various stages of anthesis (Modini et al., 1987). The eggs hatch in 2–3 days. Small, grayish, maggots establish themselves close to the developing grain and extract their food from the developing ovary of the sorghum. When feeding begins the larva turns pinkish, which deepens with growth, and attains a distinct red color at pupation. The portion of the ovary in contact with the larva shrinks, and the larva is partially enveloped in the resulting shallow, irregular depression (Gable et al., 1928). Larvae feed on the developing grain for 7–11 days (Gable et al., 1928) or 10–15 days and pupate inside the glumes (Barwad, 1981). Larvae develop more slowly in pedicellate spikelets than in sessile spikelets on sorghum panicles and such difference is useful for the determination of the development rate on midge-resistant sorghums (Franzmann, 1993). The pupa wriggles its way to the tip of the glume, and the adult comes out of the pupal case (Sharma, 2004). Adults emerge in 3–5 days (Barwad, 1981). The life cycle is completed in 14–16 days (Gable et al., 1928) or 15–20 days, and there may be 10 to 15 generations in a year if host plants are available continuously (Sharma and Vidyasagar, 1992; Sharma, 2004). The yellow sticky traps attract the greatest numbers of *S. sorghicola* (Sharma and Vidyasagar, 1992). Around 2–3% of the larvae enter into an obligate diapause of up to 3 years inside spikelets at every generation. Diapause has evolved as an adaptation to long dry spells and ends at exposure to high humidity. Exposure to continuous moisture for 10–15 days is essential for diapause termination (Passlow, 1965; Baxendale and Teetes, 1983; Sharma and Davies, 1988; Sharma, 2004; Franzmann et al., 2006).

The plant characters of sorghum, such as short, tight, light green, and hard glumes; tannin content of the grain; and rate of grain development in the initial stages, influence the ovipositioning and survival of the midge (Sharma, 1985; Sharma et al., 1990).

BEETLES

Chnootriba similis

Larval and adult feeding is typical of any phytophagous coccinellid species. They feed on the leaves, leaving one epidermis and the veins intact. This type of damage is called "windowing." Heavy populations cause a complete skeletonization and dried-up leaves.

Pale yellow eggs are laid, usually on the underside of the leaves in clusters of 20–50 eggs. Eggs are elongate oval with hexagonal sculpturing.

Larvae are covered with spines. Larvae are initially pale in color and become darker at maturity. The larva takes about 13–16 days to enter into pupation. The pupal period lasts for 6–8 days. Adults have dull reddish or orange elytra with black markings. The total life cycle is completed in 4–5 weeks in dry seasons and 7–9 weeks in rainy seasons (Schmutterer and Inyang, 1974). During winter the adults undergo diapause in Ethiopia (Beyene et al., 2009).

Chiloloba acuta

The eggs are creamy-white, oval in shape. The eggs are laid in soil and the grubs hatch in 11–16 days. Grubs feed on organic matter in the soil. There are three instars of grubs and they take about 7–8 months. The pupa takes over 3 months to reach the adult stage. Females live for 57–72 days, whereas males live for 44–57 days. The ovipositioning period lasts for 25–38 days and the female lays 23–32 eggs in its lifetime. The sex ratio of males to females was 0.92:1 and 0.67:1 in September and October 1994, respectively (Singh et al., 1997).

Myllocerus undecimpustulatus

M. undecimpustulatus is commonly called the cotton gray weevil. It is active from April to November and passes winter in the adult stage, hidden in debris. Its life cycle is completed in 6–8 weeks during the active period. The female lays on average 360 eggs over a period of 24 days. The egg period lasts for 3–5 days. *Myllocerus* species feed on rootlets (Atwal, 1976). Pupation occurs in the soil inside earthen cells and the adult emerges in a week.

The adult weevils of *Myllocerus* species feed on leaves, nibbling the leaves from the margins and eating away small patches of leaf lamina (Atwal, 1976; Ramamurthy and Ghai, 1988; Shanthichandra et al., 1990).

M. undecimpustulatus undatus is a subspecies of M. undecimpustulatus. It is found in areas with not much climatic variation. There is no inactive stage in this subspecies, unlike M. undecimpustulatus. Under laboratory conditions, the preovipositional period varied from 3 to 4 days. A single female laid 90–95 eggs during its life span and the egg period was about 7–9 days. Grubs feed on roots of the plants and pass four instars, each taking about 9–10, 11–16, 17–22, and 18–22 days, respectively. The pupal period lasts for 23–33 days. Total developmental time takes 38–56 days (Geetha and Krishna Kumar, 2013). An adult weevil of M. undecimpustulatus undatus is similar to *Artipus floridanus* and the former differs in having dark mottling on its elytra, a yellowish head, and spined femora (Ramamurthy and Ghai, 1988; Thomas and Mergendoller, 2000).

Tanymecus indicus

Adults of *T. indicus* emerge from the pupa in the soil during June and feed on the leaves of pearl millet and sorghum plants. In October, the adults move to the winter host, *Trifolium alexandrium* (Nanda and Pajni, 1990).

GRASSHOPPERS

Hieroglyphus nigrorepleptus

It is a univoltine species. There are reports of five to eight instars of the nymphal stage (Srivastava, 1956; Pradhan and Peswani, 1961; Roonwall, 1976). Eggs are laid during September–October in the soil and the nymphs emerge at the onset of monsoon showers during June–July, in India (Roonwall, 1976), and July–August in Pakistan (Sultana and Wagan, 2007). The nymphal period takes about a month.

Decticoides brevipennis

Seven nymphal instars are observed, and the adult *D. brevipennis* is a major pest of grain crops in northern Ethiopia. The time between molts for the first three instars is 10–11 days (Stretch et al., 1980).

TERMITE

Odontotermes obesus

It is a common termite especially in northern India. *O. obesus* is both a subterranean and a mound builder. Usually the colony contains 2.5% soldiers, which are sterile. Reproductively active winged males and females emerge from underground as a swarm annually, at nightfall after the first heavy rain in mid-July, and then undergo dealation. Ovipositioning was observed to occur at the rate of 273 eggs every 15 min in the rainy months and the eggs were laid in batches of 15–20 (Arora and Gilotra, 1960).

POPULATION ECOLOGY
SHOOT FLY

Population biology in terms of migration, dynamics of species combinations vis-à-vis host plants, off-season survival, residual populations, population buildup, etc., is poorly understood in *Atherigona* species.

Usually (average 1.2 eggs/plant) egg distribution follows a Poisson distribution in *A. soccata* (Raina, 1982). Low humidity affects the duration of egg development in this sorghum shoot fly. The protection against desiccation achieved by *A. soccata* eggs through both active and passive mechanisms is an indication of the adaptation of the pest to semiarid and arid environments (Delobel, 1983). Until the late 1980s, it was observed that wild gramineae plants as alternative hosts were of minor importance in the "carryover" of *A. soccata* through the off season. The irrigated sorghum grown for fodder is of major importance as a source of flies (Davies and Seshu Reddy, 1981). However, wild sorghums do carry low numbers of the sorghum shoot fly throughout the year and this assists in the maintenance of the species in the absence of cultivated sorghum (Davies and Seshu Reddy, 1981). In general, every species of *Atherigona* shows a preference toward a particular graminaceous host compared to other hosts. The numerically dominant fly reared from wild Gramineae was *Atherigona falcata*, which was recorded on 17 hosts, including sorghum. Other common species in Gramineae were *Atherigona pulla*, *Atherigona oryzae*, *Atherigona punctata*, and *Atherigona atripalpis* (Davies and Seshu Reddy, 1981). The behavior, plant injury, and biological cycle of *A. soccata* are quite consistent and uniform throughout its area of distribution (Young, 1981).

Generation Cycles in Atherigona soccata

A. soccata passes at least 11 generations in Hyderabad, India, and different generations survive on various graminaceous grasses including two seasons of cultivated sorghum (Fig. 3.16) (Kalaisekar, unpublished data).

Population Analysis of Atherigona soccata in India

Morphometric analysis using wing, head, and thorax dimensions revealed the existence of geographically varying populations. The shoot fly populations collected from different localities in India, viz., Udaipur (Rajasthan), Hyderabad (Telangana), Dharwad (Karnataka), and Kovilpatti (Tamil Nadu), morphometrically differ. The geographic distance between the locations does not contribute significantly to the variation among the populations (Fig. 3.17). The clustering patterns of different

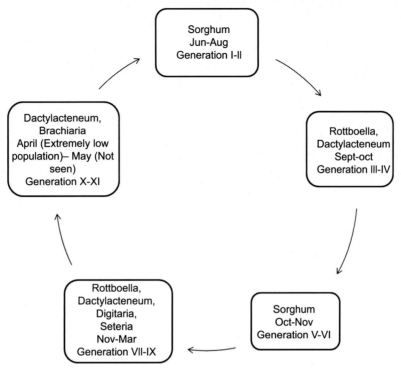

FIGURE 3.16

Generation cycle of *Atherigona soccata*.

Courtesy of A. Kalaisekar.

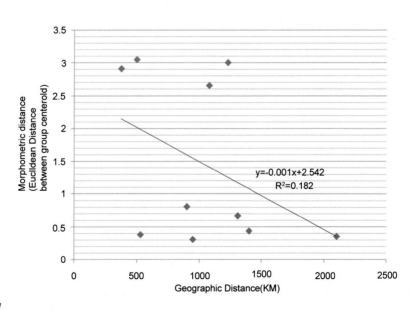

FIGURE 3.17

Effect of geographic distance between the locations on morphometric distance of populations of *Atherigona soccata*.

Courtesy of A. Kalaisekar.

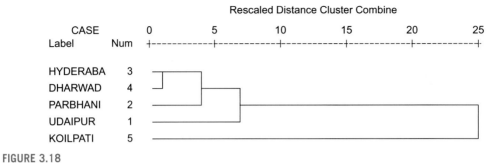

FIGURE 3.18

Clustering patterns of various geographic populations of *Atherigona soccata* in India.

Courtesy of A. Kalaisekar.

populations prove that the availability of sorghum in different seasons at a particular location influences the population variation in *A. soccata* (Fig. 3.18) (Kalaisekar, unpublished data).

Mortality Factors That Regulate Atherigona soccata *Population*

At the egg stage, egg parasitization by *Trichogrammatoidea simmondsi* was the major significant cause of mortality with 18%. After egg hatching, prolongation in duration of events, viz., first-instar larval entry into leaf sheath, pupariation, eclosion, and posteclosion, significantly contributed to the total mortality with 23%, 19%, 29%, and 20% mortality, respectively. In the case of first-instar larvae, prolonged wandering on the plant surface exposed the larvae to desiccation, which caused the mortality. In the processes of pupariation and eclosion there was a gradual decrease in the muscle contraction movement, beyond the normal time limit, which ended up in a cessation of muscular contraction and death. During posteclosion, failed or protracted ptilinal retraction caused a delay in wing expansion and tanning. Such a delay severely affected the normal development of adults and ultimately led to mortality. During the second and subsequent larval stages parasitization mainly by *Neotrichoporoides nyemitawus* resulted in statistically significant mortality (21%). In the puparium, there was a significant mortality of 13% due to parasitization by *Spalangia endius*. These results indicate that *A. soccata* was vulnerable to prolongation of time duration beyond the maximum limit and suffered maximum mortality during hatch to entry, first instar, pupariation, eclosion, and posteclosion periods. Therefore, these periods were highly critical in the life history of the fly (Kalaisekar et al., 2013). Mortality during critical phases of *A. soccata* is summarized in Table 3.2.

STEM BORERS

Stem borer moths of cereals and millets, especially the noctuoids, exhibit long-distance and short-distance dispersal flights depending upon the host plant growth stage. Long-distance flights are observed during the maturity stage of the host plant to seek new areas where younger plants are available. During such periods of long-distance dispersal flights, the moths are highly attracted toward light sources during night. After successfully establishing in a new area on younger host plants, the moths seldom disperse and fly short distances in search of mates. During this period, the moths do not show much attraction for light.

Table 3.2 Mortality During Critical Phases of *Atherigona soccata*

Phase	Time Duration Range (h)	Percentage Mortality Due to				Total Percentage Mortality
		Time Duration		Parasitization	Unknown Cause	
		< Lower Limit of Range	> Upper Limit of Range			
Egg hatching	43–74	0	3.5	18	2	23.5
First-instar larva	30–39	0	23	0	5	28
Second- to final-instar larva	120–138	0	0	21	3.3	24.3
Pupariation	3.3–4.6	0	19	0	5	24
Puparium	115–133	0	3	13	1.1	17.1
Eclosion	5.1–15	0.5	29	0	2.5	32
Posteclosion	4.4–5.1	1	20	0	5	26

C. partellus is an Asian species that invaded Africa from India sometime before 1930, when it was first recorded in Malawi (Bleszynsky, 1970; Tams, 1932; Kfir et al., 2002). *C. partellus* has been more competitive than the widespread native stem borers in Africa (Kfir et al., 2002). *C. partellus* has been displacing the native species such as *Chilo orichalcociliellus* (Ofomata et al., 1999) and *B. fusca* (Seshu Reddy, 1983) in several areas of Africa. The displacement is most evident in grain sorghum (Kfir et al., 2002).

The following are some of the important factors responsible for the competitive superiority of *C. partellus* over some native stem borers (Kfir et al., 2002).

1. *C. partellus* has a higher population growth rate and completes a generation in less time than *C. orichalcociliellus* (Kioko et al., 1995; Ofomata et al., 1999).
2. *C. partellus* terminates diapause more rapidly than *C. orichalcociliellus* (Ofomata et al., 1999) and *B. fusca* (Kfir, 1997), which may allow *C. partellus* to colonize host plants before the two native species at the beginning of growing seasons.
3. *C. partellus* has a higher rate of progeny survival than *B. fusca* and *C. orichalcociliellus* (Ofomata et al., 1999).
4. The larvae of *C. partellus*, especially neonates, disperse to greater distances than *C. orichalcociliellus*, which may allow *C. partellus* to colonize more plants than the native borer (Ofomata et al., 1999).

Interplant dispersal of neonate larvae sometimes takes place by a "ballooning" effect similar to spider mite dispersal. Larval crawling to neighboring plants is also noticed in borers.

Ovipositioning of *C. partellus* follows a random distribution and an aggregated distribution in the field. Around 25% of the plants show an aggregate distribution of eggs. About 20% of larvae hatched from an egg mass remain on the same plant 1 week after hatching. There is no density dependence of mortality of eggs and larvae due to natural enemies or other factors (Pats, 1992).

Sesamia nonagrioides glues its egg masses in a concealed manner between the leaf sheath and the stem, and the egg parasitoid *Telenomus busseolae* is able to parasitize it successfully. The colleterial glands of *S. nonagrioides* are possible sources of the host recognition kairomone for *T. busseolae* (De Santis et al., 2008).

M. separata is known to migrate hundreds of kilometers (Grist and Lever, 1969). The period before sexual maturity has been suggested as the optimum time for long-distance migration (Huang and Hao, 1966).

APHIDS

The population ecology of aphids is intertwined with that of huge numbers of natural enemies in an ecosystem. Amid heavy mortality due to predation, aphid numbers increase rapidly in sorghum fields during early summer and reach damaging numbers at the flowering stage of the plants (van Rensburg, 1973). There is infighting among natural enemies in the form of cannibalism, competition, and intraguild predation that affects the predatory efficiency (Kalaisekar, unpublished data). Then, toward crop maturity, the aphid population declines sharply owing to heavy dispersion of aphids and predation (van Rensburg, 1973).

Alates of *M. sacchari* disperse throughout the year and infest sorghum at the seedling stage itself (van Rensburg, 1973).

Corn leaf aphids, *R. maidis*, occurring on sorghum serve as an important early season food source for predaceous coccinellids. The population of these coccinellids goes up and becomes an important population regulator of greenbugs that begin to enter the sorghum field later in the season (Michels and Matis, 2008).

Four predators, viz., *Cheilomenes sexmaculata*, *Coccinella transversalis*, *Ischiodon scutellaris* and *Chrysoperla carnea*, are the major population-regulating natural enemies of *M. sacchari* in the sorghum cultivated ecosystem in south India.

MIDGES

Eulophid parasitoids, *Aprostocetus* spp., are the most important population-regulating natural enemy of the sorghum midge, *S. sorghicola* (Jotwani and Srivastava, 1976; Sharma and Davies, 1988; Nwanze et al., 1998). Weather factors, viz., temperature (25–30°C) and relative humidity (>60%), play important roles in the emergence, activity, and population buildup of the sorghum midge (Fisher and Teetes, 1982; Sharma and Vidyasagar, 1992; Kandalkar et al., 2002). Continuous drought and heavy rainfall are unfavorable to midge population growth (Jotwani, 1976). Diapausing larvae are highly sensitive to sunlight and die upon exposure to the sun (Mogal et al., 1980).

Late planting, staggered sowing (to match uneven rainfall), late flowering, low plant population, and presence of alternate hosts are the most favorable conditions for the midge population buildup (Baxendale et al., 1984; Patel and Jotwani, 1986; Sharma and Davies, 1988). Johnson grass provides an important bridging host, sustaining one or two generations of sorghum midge in Australia (Lloyd et al., 2007).

SHOOT BUG

Oligophagy in *P. maidis* might have resulted from the inclusion of introduced hosts in their feeding repertoire as a means of host expansion (Denno and Roderick, 1990). Phenolic acids interfere with the settling behavior of the shoot bug in corn (Fisk, 1980; Denno and Roderick, 1990).

High population densities, low quality of food, unstable conditions, and low temperatures favor the production of more macropterous individuals in a colony. In contrast, low population density, better quality food, and stable and favorable conditions help the production of more brachypterous adults (Fernández-Badillo, 1989). Macropterous individuals tend to disperse away in search of new areas with favorable conditions, whereas brachypterous adults divert their energy toward reproductive activities to increase the population (Fig. 3.19). Dispersal by macropterous adults to new habitats or plants helps the bugs explore host plant quality, escape overcrowding, and avoid fitness-reducing conditions (Kisimoto, 1965; Denno, 1979; Prestidge and McNeill, 1983; Denno and Roderick, 1990). The adaptive nature of polymorphic life histories existing in some plant hoppers, e.g., *P. maidis*, *Nilaparvata lugens*, and *Sogatella furcifera*, for exploiting temporary habitats or plants is known as "colonization syndrome" (Denno and Roderick, 1990). In plant hoppers such as *S. furcifera* and *N. lugens,* macrapters migrate transoceanically from China to Japan with a flight range of 600–1000 km (Kisimoto, 1972, 1976). Such a long migratory ability is mainly due to low wing loading (male: 0.0807 mg/mm^2; female: 0.0932 mg/mm^2) in *S. furcifera* (Kisimoto, 1976, 1977, 1979). Macropterous adults disperse to new habitats and initially colonize at a very

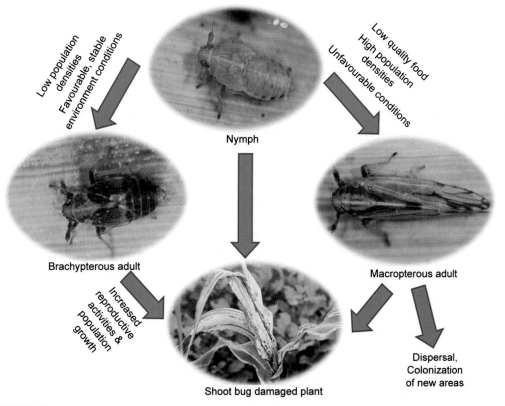

Nymph

Low population densities Favourable, stable environment conditions

Low quality food High population densities Unfavourable conditions

Brachypterous adult

Macropterous adult

Increased reproductive activities & population growth

Shoot bug damaged plant

Dispersal, Colonization of new areas

FIGURE 3.19

Population structure of *Peregrinus maidis* with reference to environmental factors.

Courtesy of A. Kalaisekar.

low population density. These macropterous adults produce individuals that are mostly brachypterous at maturity and the population rapidly grows. At this juncture, increased population density and deterioration of the food plant quality lead to the production of more and more macropterous adults (Denno and Roderick, 1990). The initial wave of immigrant macropterous populations to new plants of sorghum or corn establishes itself. Toward the maturity of the crop or if the plant is drying up because of severe infestation, or if the crop is harvested, the residual populations leave the field to settle on alternate weedy hosts. The dispersal by macropterous adults is further influenced by a variety of interactions, including the host canopy, habitat persistence, planting succession, and resource isolation (Napompeth, 1973). Tending ants have a mutualistic relationship with *P. maidis* (Khan and Rao, 1956; Fisk et al., 1981).

NILAPARVATA LUGENS

Intraspecific competition can be intense and adversely affects fitness at high population densities. High population density leads to reduced survival rates, slow growth rate, lowered fecundity, and decreased reproductive rate (Kisimoto, 1965; Denno, 1979; Kuno, 1979; Denno and Roderick, 1990).

There is an acoustic communications between individuals of the plant hoppers; males and females produce sounds by vibrating their abdomens dorsoventrally, and the vibrations are transmitted to the plant through the legs and inserted stylets. The vibrations are known to be transmitted to neighboring plants through physical contact between plants (Ichikawa, 1976; Denno and Roderick, 1990). Acoustic signals in plant hoppers facilitates male aggression, mate recognition, location, and attraction, courtship, and mate choice (Ichikawa, 1976; Denno and Roderick, 1990).

N. lugens and *P. maidis* individuals develop fastest between 25 and 28°C, and variations in temperature in either direction result in delayed development (Denno and Roderick, 1990).

CALOCORIS ANGUSTATUS

Panicle length affects the colonization of head bug populations on sorghum (Subbarayudu et al., 2010). Low temperature (<18°C) and low relative humidity (<30%) reduce *C. angustatus* density. During the postrainy season, higher temperatures (>32°C) and drought conditions have negative association with density of *C. angustatus*, *C. pallidus*, and *E. bellevoyei* (Sharma and Lopez, 1990a,b).

BLISSUS LEUCOPTERUS

The most important mortality factors of chinch bugs in the field are parasitism by the scelionids, infection by fungus, desiccation, failure to hatch, predation by *Amara* sp., and wet conditions that prevail during eclosion. Egg mortality of about 59% is the largest population-regulating factor, and a temperature of 70°F (21°C) triggers activity in the winter (Mailloux and Streu, 1981).

Unlike *B. leucopterus*, large spring dispersals are absent in *B. insularis* and they move from lawn to lawn on foot. Decreasing temperature and high rainfall are the most important weather factors that reduce the population to a great extent. The chinch bug population increases with increasing temperature in Florida, USA (Kerr, 1966).

REFERENCES

Ahee, J.E., Sinclair, B.J., Dorken, M.E., 2013. A new species of *Stenodiplosis* (Diptera: Cecidomyiidae) on florets of the invasive common reed (*Phragmites australis*) and its effects on seed production. Canadian Entomologist 145, 235–246.

Akibo-Betts, D.T., Raymundo, S.A., 1978. Aphids [*Hysteroneura setariae, Tetraneura nigriabdominalis*] as rice pests in Sierra Leone. International Rice Research Newsletter 3, 15–16.

Ammar, E.D., 1986. Ultrastructure of the salivary glands of the planthopper, *Peregrinus maidis* (Ashmead) (Homoptera : Delphacidae). International Journal of Insect Morphology and Embryology 15, 417–428.

Anderson, W.G., Heng-Moss, T.M., Baxendale, F.P., Lisa, M., Baird, L.M., Sarath, G., Higley, L., 2006. Chinch bug (Hemiptera: Blissidae) mouthpart morphology, probing frequencies, and locations on resistant and susceptible germplasm. Journal of Economic Entomology 99 (1), 212–221.

Andre, F., 1937. An undescribed chinch bug from Iowa. Iowa State College Journal of Science 11, 165–167.

Arora, G.L., Gilotra, S.K., 1960. The biology of *Odontotermes obesus* (Rambur) (Isoptera). Research Bulletin of the Panjab University of Science 10 (3–4), 247–255.

Atwal, A.S., 1976. Agricultural Pests of India and South-East Asia. Kalyani Publishers, Delhi India. x + 502 pp.

Barwad, W.L., 1981. Note on the hibernation of *Contarinia sorghicola* Coquillett in pedicellate spikelets of sorghum ear. Indian Journal of Agricultural Sciences 51, 138.

Baxendale, F.P., Teetes, G.L., Sharpe, P.J.H., 1984. Temperature-dependent model for sorghum midge (Diptera: Cecidomyiidae) spring emergence. Environmental Entomology 13, 1566–1571.

Baxendale, F.P., Teetes, G.L., 1983. Thermal requirements for emergence of overwintered sorghum midge (Diptera: Cecidomyiidae). Environmental Entomology 12, 1078–1082.

Beyene, Y., Hofsvang, T., Azerefegne, F., 2009. Phenology and migration of tef Epilachna, *Chnootriba similis* Thunberg; (Coleoptera: Coccinellidae) in Ethiopia. Journal of Entomology 6, 124–134.

Blackman, R.K., Koehler, M.M.D., Grimaila, R., Gelbart, W.M., 1989. Identification if a fully-functional hobo transposable element and its use for germ-line transformation of Drosophila. EMBO Journal 8, 211–217.

Blackman, R.L., Eastop, V.F., 1984a. Aphids on the World's Crop an Identification Guide. Wiley British Museum London, United Kingdom.

Blackman, R.L., 1987. Reproduction, cytogenetics, and development. In: Minks, A.K., Harrewijn, P. (Eds.), Aphids: their biology, natural enemies and control, vol. A. Elsevier, Amsterdam, pp. 163–195.

Blackman, R.L., Eastop, V.F., 1994a. Aphid on the World's Trees an Identification and Information Guide. CAB International Wallingford, United Kingdom.

Blackman, R.L., Eastop, V.F., 1984b. Aphids on the World's Crops. Wiley, Chichester & New York. 466 pp.

Blackman, R.L., Eastop, V.F., 1994b. Aphids on the World's Trees. CAB International, Wallingford. 987 pp. + 16 plates.

Blackman, R.L., Eastop, V.F., 2000a. Aphids on the World's Crops, second ed. Wiley, Chichester. 466 pp.

Blackman, R.L., Eastop, V.F., 2006. Aphids on the World's Herbaceous Plants and Shrubs, 2 vols. Wiley, Chichester. 1439 pp.

Blackman, R.L., Eastop, V.F., 2007a. Taxonomic issues. In: van Emden, V.F., Harrington, R. (Eds.), Aphids as Crop Pests. CAB International, Wallingford, pp. 1–29.

Blackman, R.L., Eastop, V.F., 1990. Biology and taxonomy of the aphids transmitting barley yellow dwarf virus. III. Ecology and epidemiology. In: Burnett, P.A. (Ed.), World Perspective on Barley Yellow Dwarf. CYMMYT, Mexico, D.F, Mexico, pp. 197–214.

Blackman, R.L., Eastop, V.F., 2007b. Taxonomic Issues. In: van Emden, H.F., Harrington, R. (Eds.), Aphids as Crop Pests. CABI, UK, pp. 1–29.

Blackman, R.L., Eastop, V.F., 2000b. Aphids on the world's crops: an identification and information guide. In: Aphids on the World's Crops: An Identification and Information Guide 2:x + 466 pp.; 39 pp. of ref.

Bleszynsky, S., 1970. A revision of the world species of *Chilo* Zincken (Lepidoptera: Pyralidae). Bulletin of the British Museum (Natural History) Entomology 25, 101–195.

Bosque-Pérez, N.A., Dabrowski, Z.T., 1989. Mass rearing of the maize stem borers *Sesamia calamistis* and *Eldana saccharina* at IITA. In: Toward Insect Resistant Maize for the Third World: Proceedings of the International Symposium on Methodologies for Developing Host Plant Resistance to Maize Insects. Centro Internacional de Mejoramiento de Maíz y Trigo., Mexico, D.F, pp. 22–26.

Bosque-Perez, N.A., Alam, M.S., 1992. Mass Rearing of Cicadulina Leafhoppers to Screen for Maize Streak Virus Resistance. A Manual. International Institute of Tropical Agriculture, Ibadan, Nigeria. 22 pp.

Bosque-Perez, N.A., Schulthess, F., 1998. Maize: West and Central Africa. In: Polaszek, A. (Ed.), African Cereal Stem Borer: Economic Importance, Taxonomy, Natural Enemies and Control. CAB International, England, 530 p.

Breniere, J., Rodriguez, H., Ranaivosoa, H., 1962. Un ennemi de riz 'a Madagascar, *Maliarpha separatella* Rag. ou boreur blanc. *Agronomia Tropical* 17, 223–302.

Brewer, M.J., Elliot, N.C., 2004. Biological control of cereal aphids in North America and mediating effects of host plant and habitat manipulations. Annual Review of Entomology 49, 219–242.

Brożek, J., Bourgoin, T., Szwedo, J., 2006. The interlocking mechanism of maxillae and mandibles in Fulgoroidea (Insecta: Hemiptera: Fulgoromorpha). Polish Journal of Entomology 75, 239–253.

Bruno, L., Irene, M.A., 2013. The digestive tract of *Drosophila melanogaster*. Annual Review of Genetics 47, 377–404.

Bullock, J.A., 1970. The distribution of attack by a shoot-fly larva. Journal of Applied Ecology 7, 463–479.

Burd, J.D., Porter, D.R., 2006. Biotypic diversity in greenbug (Hemiptera: Aphididae): characterizing new virulence and host associations. Journal of Economic Entomology 99, 959–965.

Busey, P., Zaenker, E.I., 1992. Resistance bioassay from southern chinch bug (Heteroptera: Lygaeidae) excreta. Journal of Economic Entomology 85, 2032–2038.

Catindig, J.D., Barrion, A.T., Litsinger, J.A., 1996. Plant host range and life history of the corn delphacid, *Peregrinus maidis* (Ashmead) (Hemiptera: Delphacidae). Asia Life Sciences (Philippines) 5, 35–46.

Chakrabarti, S., Raychaudhuri, D.N., 1978. Callipterine aphids (Homoptera: Aphididae) of north-east India and Bhutan. Proceedings of the Zoological Society (Calcutta) 28, 71–101.

Chang, N.T., 1981. Resistance of some grain sorghum cultivars to sorghum aphid injury. Plant Protection Bulletin, Taiwan 23 (1), 35–41.

Chelliah, S., Basheer, M., 1965. Biological studies of *Peregrinus maidis* (Ashmead) (Araeopidae: Homoptera) on sorghum. Indian Journal of Entomology 27, 466–471.

Dabrowski, Z.T., Patel, N.Y., 1981. Investigations on physiological components of *Atherigona soccata* larvae and their interaction with sorghum—I. Larval enzymes. International Journal of Tropical Insect Science 2 (1–2), 73–76.

Dahms, R.G., 1948. Effect of different varieties and ages of sorghum on the biology of the chinch bug. Journal of Agricultural Research 76, 271–288.

Dahms, R.G., Snelling, R.O., Fenton, F.A., 1936. Effect of different varieties of sorghum on biology of the chinch bug. Journal of American Society of Agronomy 28, 160–161.

Damte, T., Pendleton, B.B., Almas, L.K., 2009. Cost–benefit analysis of sorghum midge, *Stenodiplosis sorghicola* (Coquillett)-resistant sorghum hybrid research and development in Texas. Southwestern Entomologist 34, 395–405.

David, S.K., 1976. A taxonomic review of *Macrosiphum* in India. Oriental Insects 9, 461–493.

David, S.K., Sandhu, G.S., 1976. New oviparous morph on *Melanaphis sacchari* (Zehntner) on sorghum. *Entomological Records* 88, 28–29.

Davies, J.C., Seshu Reddy, K.V., 1981. Shootfly species and their graminaceous hosts in Andhra Pradesh, India. Insect Science and Its Application 2, 33–37.

Delobel, A., 1983. Etude des facteurs déterminant l'abondance des populations de la mouche du sorgho, *Atherigona soccata* Rondani (Diptères, Muscidae) Thèse de Doctorat d'Etat, Université de Paris Sud, Centre d'Orsay. ORSTOM, Paris. 127 pp.

Delobel, A.G.L., 1982. Effects of sorghum density on oviposition and survival of the sorghum shoot fly, *Atherigona soccata*. Entomologia Experimentalis et Applicata 31, 170–174.

Delobel, A.G.L., 1981. The distribution of the eggs of the sorghum shootfly, *Atherigona soccata* Rondani (Diptera: muscidae). Insect Science and Its Applications 2, 63–66.

Denno, R.F., 1979. The relation between habitat stability and the migration tactics of planthoppers. Miscellaneous Publications of Entomological Society of America 11, 41–49.

Denno, R.F., Roderick, G.K., 1990. Population biology of planthoppers. Annual Review of Entomology 35, 489–520.

De Santis, F., Conti, E., Romani, R., Salerno, G.A., Parillo, F., Bin, F., 2008. Colleterial glands of *Sesamia nonagrioides* as a source of the host-recognition kairomone for the egg parasitoid, *Telenomus busseolae*. Physiological Entomology 33, 7–16.

Duale, A.H., Nwanze, K.F., 1999. Incidence and distribution in sorghum of the spotted stem borer *Chilo partellus* and associated natural enemies in farmers' fields in Andhra Pradesh and Maharashtra states. International Journal of Pest Management 45 (1), 3–7 25 ref.

Eastop, V.F., 1966. A taxonomic study of Australian Aphidoidea. Australian Journal of Zoology 14, 399–592.

Estela, L.A., Jose, L.S., 2010. Insect fat body: energy, metabolism, and regulation. Annual Review of Entomology 55, 207–225.

Fernández-Badillo, A., 1988. Descripcion e identificacion de los huevos e instares de la chicharrita del maiz, *Peregrinus maidis* (Homoptera: Delphacidae). Memoria de la Sociedad de Ciencias Naturales La Salle 48, 117–125.

Fernández-Badillo, A., 1989. El nombre valido de la chicharrita del maiz, Peregrinus maidis (Homoptera: Delphacidae). Boletin de Entomologia Venezolana 5 (4), 37–38.

Ferrar, P., 1979. Cocoon formation by muscidae (Diptera). Journal of Australian Entomological Society Melbourne 19, 171–174.

Fisher, R.W., Teetes, G.L., 1982. Effects of moisture on sorghum midge (Diptera: Cecidomyiidae) emergence. Environmental Entomology 11, 946–948.

Fisk, J., 1980. Effects of HCN, phenolic acids and related compounds in Sorghum bicolor on the feeding behaviour of the planthopper, *Peregrinus maidis*. Entomologia Expermentalis et Applicata 27 (21), 1–22.

Fisk, J., Bernays, E.A., Chapman, R.F., Woodhead, S., 1981. Report O F Studies on the Feeding Biology of *Peregrinus Maidis*. Cent. For Overseas Pest Res., Int. Crops Res. Inst. for Semi-arid Trop., Core Programme Project 27. COPR, London.

Fitch, A., 1855. Chinch bug. Transactions of the New York State Agricultural Society 15, 509–527.

Foott, W.H., 1977. Biology of the corn leaf aphid, *Rhopalosiphum maidis* (Homoptera: Aphididae), in southwestern Ontario. Canadian Entomologist 109 (8), 1129–1135.

Franzmann, B.A., 1993. Ovipositional antixenosis to *Contarinia sorghicola* (Coquillett) (Diptera: Cecidomyiidae) in grain sorghum. Journal of Australian Entomology Society 32, 59–64.

Franzmann, B.A., Lloyd, R.J., Zalucki, M.P., 2006. Effect of soil burial depth and wetting on mortality of diapausing larvae and patterns of post-diapause adult emergence of sorghum midge, *Stenodiplosis sorghicola* (Coquillett) (Diptera: Cecidomyiidae). Australian Journal of Entomology 45, 192–197.

Gable, C.H., Baker, W.A., Woodruff, L.C., 1928. The Sorghum Midge, with Suggestions for Control. Farmer's Bulletin No.1566. USDA, Washington DC, p. 12.

Gadiyappanavar, R.D., ChannaBasavanna, G.P., 1973. Bionomics of the ragi (*Eleusine coracana*) root aphid, *Tetraneura nigriabdominalis* (Sasaki) (Homoptera: Pemphigidae). Mysore Journal of Agricultural Sciences 7 (3), 436–444.

Gama, Z.P., Morlacchi, P., Giorgi, A., Lozzia, G.C., Baumgaertner, J., 2010. Towards a better understanding of the dynamics of *Aphis spiraecola* Patch (Homoptera: Aphididae) populations in commercial alpine yarrow fields. Journal of Entomological and Acarological Research 42, 103–116.

Geetha, G.T., Krishna Kumar, N.K., 2013. Laboratory rearing of *Myllocerus undecimpustulatus undatus* Marshall in laboratory with special reference to its biology. Entomology, Ornithology, and Herpetology 2, 2.

Ghosh, L.K., 2008. Handbook on Hemipteran Pests in India. Zool. Survey of India, Kolkata, India, p. 453.

Goftishu, et al., 2004a. Journal of Pest Science 82 (1), 67–71.

Goftishu, M., Tefera, T., Getu, E., 2004b. Biology of barley shoot fly *Delia flavibasis* Stein (Diptera: anthomyiidae) on resistant and susceptible barley cultivars. International Journal of Pest Management 50 (1), 29–34.

Grist, D.H., Lever, R.J.A.W., 1969. Pests of Rice. Longman, London, UK.

Grzywacz, A., Pape, T., Hudson, W.G., Gomez, S., 2013. Morphology of immature stages of *Atherigona reversura* (Diptera: muscidae), with notes on the recent invasion of North America. Journal of Natural History 47 (15–16), 1055–1067.

Gullan, P.J., Cranston, P.S., 2010. Insects: An Outline of Entomology. Chapman and Hall, London.

Hales, D., Foottit, R.G., Eric Maw, E., 2010. Morphometric studies of the genus *Sitobion* Mordvilko 1914 in Australia (Hemiptera:Aphididae). Australian Journal of Entomology 49, 341–353.

Hans van, D., Sofie, R., 2010. Egg spreading in the ant-parasitic butterfly, *Maculinea alcon*: from individual behaviour to egg distribution pattern. Animal Behaviour 80, 621–627.

Hansen, A.A., 1923. Wild corn, a serious weed in Indiana. Proceedings of the Indian Academy of Science 33, 295–296.

Herold, F., Munz, K., 1967. Ultrastructure of spermatozoa of *Peregrinus maidis*. Zeitschrift fur Zellforschung und mikroskopische Anatomie 83, 364.

Harris, K.M., 1990. Biology of *Chilo* species. Insect Science and Its Application 11 (4/5), 467–477.

Harris, K.M., 1989. Bioecology of sorghum stemborers. In: International Workshop on Sorghum Stem Borers, 17–20 November 1987. ICRISAT center, Patencheru, A.P. 502 324, India, pp. 63–71.

Harris, K.M., 1962. Lepidopterous stem borers of cereals in Nigeria. Bulletin of Entomological Research 53, 139–171.

Harris, K.M., 1976. The sorghum midge. Annals of Applied Biology 64, 114–118.

Harris, K.M., Nwanze, K.F., 1992. *Busseola fusca* (Fuller), the African Maize Stalk Borer: A Handbook of Information. Information Bulletin No. 33. Patancheru, A.P. 502 324, India: International Crops Research Institute for the Semi-Arid Tropics, and Wallingford UK: CAB International. 92 pp.

Harris, M.O., Stuart, J.J., Mohan, M., Nair, S., Lamb, R.J., Rohfritsch, O., 2003. Grasses and gall midges: plant defense and insect adaptation. Annual Review of Entomology 48, 549–577.

Harvey, T.L., Hackerott, H.L., 1969. Recognition of a greenbug biotype injurious to sorghum. Journal of Economic Entomology 62, 776–779.

Hayes, W.P., 1922. Observations of insects attacking sorghums. Journal of Economic Entomology 15, 349–356.

Heinrichs, E.A., Barrion, A.T., 2004. In: Heinrichs, E.A., Barrion, A.T., Hettel, G.P. (Eds.), Rice-feeding Insects and Selected Natural Enemies in West Africa: Biology, Ecology and Identification. International Rice Research Institute (IRRI), Manila, Philippines. vi + 242 pp.

Heng-Moss, T.M., Baxendale, F.P., Riordan, T.P., Young, L., Lee, K., 2003. Chinch bug resistant buffalo-grass: an investigation of tolerance, antixenosis, and antibiosis. Journal of Economic Entomology 96, 1942–1951.

Hille Ris Lambers, D., 1966. Polymorphism in Aphididae. Annual Review of Entomology 11, 47–78.

Huang, K.H., Hao, W.W., 1966. Studies on the flight of the armyworm moth *Leucania separata*. 1. Flight duration and wingbeat frequency. Acta Entomologica Sinica 15, 96–104.

Ichikawa, T., 1976. Mutual communication by substrate vibrations in the mating behavior of planthoppers (Homoptera: Delphacidae). Applied Entomology and Zoology 1, 8–21.

Ingram, J.W., Summers, E.M., 1938. Transmission of sugarcane mosaic by the greenbug (Toxoptera graminum Rond.). Journal of Agricultural Research 56 (7).

Ismail, A.M., Hall, A.E., Close, T.J., 1997. Chilling tolerance during emergence of cowpea associated with a dehydrin and slow electrolyte leakage. Crop Science 37, 1270–1277.

Jalali, S.K., Singh, S.P., 2004. Bio-ecology of *Chilo partellus* (Swinhoe) (Lepidoptera:Pyralidae) and evaluation of its natural enemies-a review. Agricultural Reviews 24 (2), 79–100.

Jones, C.J., Milne, D.E., Patterson, R.S., Schreiber, E.T., Milio, J.A., 1992. Nectar feeding by *Stomoxys calcitrans* (Diptera: Muscidae): effects on reproduction and survival. Environmental Entomology 21, 141–147.

Jones, D.B., Giles, K.L., Elliott, N.C., 2008. Supercooling points of *Lysiphlebus testaceipes* and its host *Schizaphis graminum*. Environmental Entomology 37, 1063–1068.

Joshi, S., Viraktamath, C.A., 2004. The sugarcane woolly aphid, *Ceratovacuna lanigera* Zehnter (Hemiptera: Aphididae) its biology, pest status and control. Current Science 87, 307–316.

Jotwani, M.G., 1976. Host plant resistance with special reference to sorghum. Proceedings of the National Academy of Scienc India 46 (1–2), 42–48.

Jotwani, M.G., Srivastava, K.P., 1976. Efficacy of granular insecticides applied as whorl treatment for the control of sorghum stem borer *Chilo partellus* (Swinhoe). Entomologists' Newsletter 6 (8–9), 50–52.

Kalaisekar, A., Patil, J.V., Shyam Prasad, G., Bhagwat, V.R., Padmaja, P.G., Subbarayudu, B., Srinivasa Babu, K., Rahman, Z., 2013. Time-lapse tracing of biological events in an endophytic schizophoran fly, *Atherigona soccata* Rondani (Diptera: muscidae). Current Science 105 (5), 695–701.

Kandalkar, H.G., Men, U.B., Kadam, P., Gaikwad, M., 2002. Effect of abiotic factors on population dynamics of sorghum *midge, Contarinia sorghicola* Coq. Journal of Entomological Research 26 (2), 113–115.

Kerr, S.H., 1966. Biology of the lawn chinch bug. The Florida Entomologist 49, 9–18.

Kfir, R., 1988. Hibernation by the lepidopteran stalk borers, *Busseola fusca* and *Chilo partellus* on grain sorghum. Entomologia Experimentalis et Applicata 48, 31–36.

Kfir, R., 1991. Duration of diapause in the stem borers, *Busseola fusca* and Chilo partellus. Entomologia Experimentalis et Applicata 61 (3), 265–270.

Kfir, R., 1998. Maize and grain sorghum: Southern Africa, pp. 29–37. In: Polaszek, A. (Ed.), African Cereal Stem Borer: Economic Importance, Taxonomy, Natural Enemies and Control. CAB International, England, 530 p.

Kfir, R., 1997. Competitive displacement of *Busseola fusca* (Lepidoptera: Noctuidae) by *Chilo partellus* (Lepidoptera: Pyralidae). Annals of the Entomological Society of America 90, 620–624.

Kfir, R., Overholt, W.A., Khan, Z.R., Polaszek, A., 2002. Biology and management of economically important lepidopteran cereal stem borers in Africa. Annual Review of Entomology 47, 701–731.

Khan, M.Q., Rao, A.S., 1956. The influence of the blank ant (*Componotus compressus* F.) on the incidence of two homopteran crop pests. Indian Journal of Entomology 18 (1), 99–200.

Kim, H., Lee, W., Lee, S., 2006. Three new records of the genus *Aphis* (Hemiptera: Aphididae) from Korea. Journal of Asia-Pacific Entomology 9 (4), 302.

Kioko, E.N., Overholt, W.A., Mueke, J.M., 1995. Larval development in *Chilo orichacociliellus* and *Chilo partellus*: a comparative study in the laboratory. In: Proc. Meet. Sci. Conf. Afr. Assoc. Insect Sci., 10th, Mombasa, 1993. Afr. Assoc. Insect Sci., Nairobi, pp. 191–198.

Kisimoto, R., 1965. Studies on polymorphism and its role playing in the population growth of the brown planthopper, *Nilaparvata lugens* Stal. Bulletin of Shikoku Agriculture Experimental Station 13, 1–106.

Kisimoto, R., 1972. Long distance migration of planthoppers. Japan Pest Information 10, 115–116.

Kisimoto, R., 1976. Synoptic weather conditions inducing long-distance immigration of planthoppers, *Sogarella fitrcifera* Horvath and *Nilapanota lugens* Stal. Ecological Entomology 1, 95–109.

Kisimoto, R., 1977. Brown planthopper migration. In: IRRI (Ed.), Brown planthopper symposium, 18–22, April 1977. International Rice Research Institute. IRRI, Manila, 15 p.

Kisimoto, R., 1979. Brown Planthopper: Threat to Rice Production in Asia. IRRI, Manila, pp. 113–124.

Kundu, G.G., Kishore, P., 1970. Biology of the sorghum shoot fly, *Atherigona soccata* Rondani (Anthomyiidae: Diptera). Indian Journal of Entomology 32, 215–217.

Kuno, E., 1979. Ecology of the brown planthopper in temperate regions. In: International Rice Research Institute. Brown Planthopper: Threat To Rice Production in Asia. Los Banos, PhilippinesInt. Rice Res. Lnst, vol. 107, pp. 45–60.

Lee, C.Y., Lim, C.Y., Darah, I., 2002. Survey on structure-infesting ants (Hymenoptera: Formicidae) in food preparative outlets. Tropical Biomedicine 19, 21–26.

Leonard, D.E., 1966. Biosystematics of the "leucopterus complex" of the genus Blissus (Heteroptera: Lygaeidae). Bulletin of the Connecticut Agricultural Experiment Station 677, 47.

Lloyd, R.J., Franzmann, B.A., Zalucki, M.P., 2007. Seasonal incidence of *Stenodiplosis sorghicola* (Coquillett) (Diptera: Cecidomyiidae) and its parasitoids on *Sorghum halepense* (L.) pers. In south-eastern Queensland, Australia. Australian Journal of Entomology 46, 23–28.

Lockey, K.H., 1988. Lipids of the insect cuticle-origin, composition and function. Comparative Biochemistry and Physiology Part B 89, 595–645.

Louis, K.P., Ramasamy, P., 2008. Leaf-footed Bug, *Leptoglossus phyllopus* (Hemiptera: Coreidae), as a potential vector of sorghum fungal pathogens. Southwestern Entomologist 33 (2), 161–164.

Mailloux, G., Streu, H.T., 1981. Population biology of the hairy chinch bug (Blissus leucopterus hirtus Montandon: Hemiptera: Lygaeidae). Annals of the Entomological Society of Quebec 26, 51–90.

Margaret, G.J., 1969. Oviposition of frit fly (*Oscinella frit* L.) on oat seedlings and subsequent larval development. Journal of Applied Ecology 6, 411–424.

Mbapila, J.C., Overholt, W.A., Kayumbo, H.Y., 2002. Comparative development and population growth of an exotic stemborer, *Chilo partellus* (Swinhoe), and an ecologically similar congener, *C. orichalcociliellus* (Strand) (Lepidoptera: Crambidae). International Journal of Tropical Insect Science 22, 21–27.

McDonald, R.S., Borden, J.H., 1996. Courtship behavior and discrimination between potential mates by male Delia antiqua (Diptera: Anthomyiidae). Journal of Insect Behaviour 9, 871–885.

Meksongsee, B., Chawanapong, M., 1985. Sorghum insect pests in south east Asia. In: Proceedings of the International Sorghum Entomology Workshop, July 15–24, 1984, Texas A&M University, College Station, Texas, USA. International Crops Research Institute for the Semi-Arid Tropics (ICRISAT), Patancheru, Andhra Pradesh 502 324, India, pp. 57–64.

Meksongsee, B., Kongkanjana, A., Sangkasuwan, U., Young, W.R., 1978. Longevity and oviposition of sorghum shoot fly adults on different diets. Annals of the Entomological Society of America 71, 852–853.

Meksongsee, B., Chawanapong, M., Sangkasuwan, U., Poonyathaworn, P., 1981. The biology and control of the sorghum shoot fly, *Atherigona soccata* Rondani, in Thailand. Insect Science and its Application 2, 111–116.

Michael, F.B., 1994. The significance of egg size variation in butterflies in relation to hostplant quality. OIKOS 71, 119–129.

Michels Jr., G.J., 1986. Graminaceous North American host plants of the greenbug with notes on biotypes. Southwestern Entomologist 11, 55–66.

Michels, G.J., Matis, J.H., 2008. Corn leaf aphid, *Rhopalosiphum maidis* (Hemiptera: Aphididae), is a key to greenbug, *Schizaphis graminum* (Hemiptera: Aphididae), biological control in grain sorghum, Sorghum bicolor. Europen Journal of Entomology 105, 513–520.

Modini, M.P., Page, F.D., Franzmann, B.A., 1987. Diurnal oviposition activity in grain sorghum by *Contarinia sorghicola* (Coquillett) (Diptera: Cecidomyiidae). Journal of Australian Entomology Society 26 (4), 293–294.

Mogal, B.H., Mali, A.R., Rajput, S.G., Pawar, K.L., 1980. Chemical control of sorghum midge (Contarinia sorghicola Coq.). Journal of Maharashtra Agricultural Universities 5, 5–9.

Mordvilko, A., 1935. Die Blattlause mit unvollstandigem Generationszyklus und ihre Entstehung. Ergebnisse Fortschritte der Zoologie 8, 36–328.

Muhammad, L., Underwood, E., 2004. The maize agricultural context in Kenya. In: Hillbeck, A., Andow, D.A. (Eds.), Environmental Risk Assessment of Genetically Modified Organisms. A Case Study of Bt Maize in Kenya, vol. 1. CAB International, Wallingford, UK. 281 pp.

Murphy, H.C., 1959. The epidemic of barley yellow dwarf on oats in 1959: introduction. Plant Disease Report (Suppl. 262), 316.

Nagaraj, N.J., Reese, J.C., Kirkham, M.B., Kofoid, K., Campbell, L.R., Loughin, T., 2002. Effect of greenbug, *Schizaphis graminum* (Rondani) (Homoptera: Aphididae), biotype K on chlorophyll content and photosynthetic rate of tolerant and susceptible sorghum hybrids. Journal of the Kansas Entomological Society 75 (4), 299–307.

Nanda, N., Pajni, H.R., 1990. Notes on the life history and habits of *Tanymecus indicus* Fst. (Coleoptera: Curculionidae). Research Bulletin of the Panjab University of Science 411 (4), 61–65.

Napompeth, B., 1973. Ecology and Population Dynamics of the Corn Planthopper, *Peregrinus Maidis* (Ashmead) (Homoptera: Delphacidae) in Hawaii (Ph.D. thesis). University of Hawaii, Honolulu, Hawaii. 257 pp.

Nault, L.R., Bradley, R.H.E., 1969. Acquisition of maize dwarf mosaic virus by the greenbug, *Schizaphis graminum*. Annals of the Entomological Society of America 62, 403–406.

Ngi-Song, A.J., Overholt, W.A., Njagi, P.G.N., Dicke, M., Ayertey, J.N., Lwande, W., 1996. Volatile infochemicals used in host and host habitat location by *Cotesia flavipes* Cameron and *Cotesia sesamiae* (Cameron) (Hymenoptera: Braconidae), larval parasitoids of stemborers on Graminae. Journal of Chemical Ecology 22, 307–322.

Nuessly, G.S., 2011. Yellow Sugarcane Aphid, *Sipha flava* (Forbes) (Insecta:Hemiptera: Aphididae). http://entomology.ifas.ufl.edu/creatures.

Nuessly, G.S., Nagata, R.T., 2005. Greenbug, Schizaphis graminum (Rondani) (Insecta: Hemiptera: Aphididae). http://entomology.ifas.ufl.edu/creatures.

Nwanze, K.F., Reddy, Y.V.R., Nwilene, F.E., Kausalya, K.G., Reddy, D.D.R., 1998. Tritrophic interactions in sorghum, midge (*Stenodiplosis sorghicola*) and its parasitoid (*Aprostocetus* spp.). Crop Protection 17, 165–169.

Nye, I.W.B., 1959. The distribution of shoot-fly larvae (Diptera, Acalypterae) within Pasture grasses and cereals in England. Bulletin of Entomological Research 50, 53–62.

Nye, I.W.B., 1960. The Insect Pests of Graminaceous Crops in East Africa. Colonial Research Study No 3 1. Her Majesty' s Stationer y Office, London, UK. 48 pp.

Ofomata, V.C., Overholt, W.A., Lux, S.A., Van Huis, A., Egwuatu, R.I., 2000. Comparative studies on the fecundity, egg survival, larval feeding, and development of *Chilo partellus* and *Chilo orichalcociliellus* (Lepidoptera: Crambidae) on five grasses. Annals of the Entomological Society of America 93 (3), 492–499.

Ofomata, V.C., Overholt, W.A., Egwuatu, R.I., 1999. Diapause termination of *Chilo partellus* (Swinhoe) and *Chilo orichalcociliellus* Strand (Lepidoptera: Pyralidae). Insect Science and Its Application 19, 187–191.

Ogwaro, K., 1978. Observations on longevity and fecundity of the sorghum shoot fly, *atherigona soccata* (diptera; anthomyiidae). Entomologia Experimentalis et Applicata 23, 131–138.

Ogwaro, K., Kokwaro, E.D., 1981a. Morphological observations on sensory structures on the ovipositor and tarsi of the female and on the head capsule of the larva of the sorghum shootfly, Atherigona soccata Rondani. International Journal of Tropical Insect Science 2 (1–2), 25–32.

Ogwaro, K., Kokwaro, E.D., 1981b. Development and morphology of the immature stages of the sorghum shoot fly, *Atherigona soccata* Rondani. Insect Science and Its Application 1, 365–372.

Okoth, V.A.O., Dabrowski, Z.T., HFvan, E., 1987. Comparative biology of some Cicadulina species and populations from various climatic zones in Nigeria (Hemiptera: Cicadellidae). Bulletin of Entomological Research 77 (1), 1–8.

Oraze, M.J., Grigarick, A.A., 1992. Biological control of Ducksalad (*Heteranthera limosa*) by the Waterlily aphid (*Rhopalosiphum nymphaeae*) in rice (*Oryza sativa*). Weed Science 40 (2), 333–336.

Ortego, J., 1998. Pulgones de la Patagonia Argentina con la descripción de *Aphis intrusa*. Revista de la Facultad de Agronomia (La Plata) 102, 59–80.

Osborn, H., 1888. The chinch bug in Iowa. Iowa Agriculture College Department of Entomology Bulletin 13 pp.

Palma, M.F.M., 1988. Incidence and distribution of *Contarinia sorghicola* (CoquiUeu, 1899) (Diptera: Cecidomyiidae) in sorghum at different times of the year. Anais da Sociedade Entomologica do Brasil 17 (2), 507–518.

Passlow, T., 1965. Bionomics of sorghum midge (*Contarlnia sorghicola* (Coq.)) in Queensland, with particular reference to diapause. Queensland Journal of Agricultural and Animal Science 22 (2), 150–167.

Patel, J.R., Jotwani, M.G., 1986. Effect of ecological factors on incidence and damage by sorghum midge, Co1ltarinia sorghico/a. Indian Journal of Entomology 48 (2), 220–222.

Pats, P., 1992. Reproductive Biology of the Cereal Stemborers Chilo Partellus (Ph.D. thesis). Swed. Univ. Agric. Sci, Uppsala, Swed. 97 pp.

Petralia, R.S., Wuensche, A.L., Teetes, G.L., Sorensen, A.A., 1979. External morphology of the mouthparts of larvae of sorghum midge, Contarinia sorghicola. Annals of Entomological Society of America 72 (6), 850–855.

Pol, J.J., Sakate, P.J., 2002. Lipase Activity during Larval Development of *Chilo Partellus* (Swinhoe). Modern Trends in Entomological Researches in India, 55.

Ponnaiya, B.W.X., 1951. Studies in the genus *Sorghum*. I. Field observations on sorghum resistance to the insect pest, *Atherigona indica* M. Madras Agricultural Journal 21, 96–117.

Ponsen, M.B., 1977. Anatomy of an aphid vector: *Mysus persicae*. Chapter 2. In: Harris, K.F., Maramorosch, K. (Eds.), Aphis as Virus Vectors. Academic Press, new York.

Pradhan, S., Peswani, K.M., 1961. Studies on the ecology and control of *Hieroglyphus nigrorepletus* Boliver (Phadka). Indian Journal of Entomology 23, 79–106.

Prestidge, R.A., McNeill, S., 1983. Auchenorrhyncha-host plant interactions:leafhoppers and grasses. Ecological Entomology 8, 331–339.

Pucci, C., Forcina, A., 1984. Morphological differences between the eggs of *Sesamia cretica* (Led.) and *S. nonagrioides* (Lef.) (Lepidoptera: Noctuidae). International Journal of Insect Morphology and Embryology 13, 249–253.

Puterka, G.J., Peters, D.C., 1990. Sexual reproduction and inheritance of virulence in the greenbug *Schizaphis graminum* (Rondani), pp. 289–318. In: Campbell, R.K., Eikenbary, R.D. (Eds.), Aphid-plant Genotype Interactions. Elsevier, Amsterdam.

Rahman, M., Khalequzzaman, M., 2004. Temperature requirements for the development and survival of rice stemborers in laboratory conditions. Insect Science 11 (1), 47–60.

Raina, A.K., 1981. Deterrence of repeated oviposition in sorghum shootfly *Atherigona soccata*. Journal of Chemical Ecology 7, 785–790.

Raina, A.K., 1982. Fecundity and oviposition behaviour of the sorghum shootfly, *Atherigona soccata*. Entomologia Experimentalis et Applicata 31, 381–385.

Rajasekhar, P., 1989. Studies on the Population Dynamics of Major Pests of Sorghum and Bioecology and Crop Loss Assessement Due to the Shoot Bug, Peregrines Maidis Ashmeas (M.Sc. thesis). Andhra Pradesh agricultural university, Hyderabad, Andhra Pradesh, India. 230 p.

Ramamurthy, V.V., Ghai, S., 1988. A study on the genus *Myllocerus* (Coleoptera: Curculionidae). Oriental Insects 22, 377–500 (Requested).

Ramm, C., Wayadande, A., Baird, L., Nandakumar, R., Madayiputhiya, N., Amundsen, K., Donze-Reiner, T., Baxendale, F., Sarath, G., Heng-Moss, T., 2015. Morphology and Proteome characterization of the salivary glands of the western chinch bug (Hemiptera: Blissidae). Journal of Economic Entomology 108 (4), 2055–2064.

Ranade, D.R., 1965. Anatomy of the tracheal system of the larva of Musca domestica nebula Fabr. (Diptera: Muscidae). Indian Journal of Entomology 27 (2), 172–181.

Raychaudhuri, D.N. (Ed.), 1980. Aphids of North-East India and Bhutan. Zoological Society, Calcutta. 521 pp.

Rioja, T.M., Vargas, H.E., Bobadilla, D.E., 2006. Biology and natural enemies of *Peregrinus maidis* (Ashmead) (Hemiptera: Delphacidae) in the Azapa valley. IDESIA (Chile) 24, 41–48.

Roonwall, M.L., 1976. Ecology and biology of the grasshoppers *Hieroglyphus Nigrorepletus* Bolivar (Orthoptera: Acrididae) distribution, economic importance, life history, color forms and problems of control. Zeitschrift fuer Angewandte Zoologie Berlin 63: 307–323.

Rose, D.J.W., 1973a. Distances flown by Cicadulina spp. (Hem., Cicadellidae) in relation to distribution of maize streak disease in Rhodesia. Bulletin of Entomological Research 62 (3), 497–505.

Rose, D.J.W., 1973b. Field studies in Rhodesia on Cicadulina spp. (Hem., Cicadellidae), vectors of maize streak disease. Bulletin of Entomological Research 62 (3), 477–495.

Rose, D.J.W., 1973c. Laboratory observations on the biology of Cicadulina spp. (Hom., Cicadellidae), with particular reference to the effects of temperature. Bulletin of Entomological Research 62 (3), 471–476.

Rothschild, G.H.L., 1971. The biology and ecology of rice stem borers in Sarawak (Malaysian Borneo). Journal of Applied Ecology 8, 287–322.

Royer, T.A., Pendleton, B.B., Elliott, N.C., Giles, K.L., 2015. Greenbug (Hemiptera: Aphididae) biology, ecology, and management in wheat and sorghum. Journal of Integrated Pest Management 6 (1), 19. http://dx.doi.org/10.1093/jipm/pmv018.

Sakate, P.J., Pol, J.J., 2002. Lipase activity in the fat body of *Chilo partellus* during larval growth and metamorphosis. ENTOMON 27 (2), 147–152.

Sallam, M.N., Overholt, W.A., Kairu, E., 1999. Comparative evaluation of Cotesia flavipes and C. sesamiae (Hymenoptera: Braconidae) for the management of Chilo partellus (Lepidoptera: Pyralidae) in Kenya. Bulletin of Entomological Research 89 (02), 185–191.

Sarup, P., Panwar, V.P.S., 1987. Preferential plant age and site for oviposition by *Atherigona naqvii* Steyskal and *Atherigona soccata* Rondani in spring sown maize. J. Entomol. Res. 11 (2), 150–154.

Scheltes, P., 1978a. Ecological and Physiological Aspect of Aestivation-diapause in the Larvae of Two Pyralid Stalk Borers of Maize in Kenya (Ph.D. thesis). Landbouwhogeschool, Wageningen. 110 pp.

Scheltes, P., 1978b. The condition of the host plant during aestivation-diapause of the stalk borers *Chilo partellus* and *Chilo orichalcociliella* (Lepidoptera: Pyralidae) in Kenya. Entomologia Experimentalis et Applicata 24, 679–688.

Schmutterer, G., Inyang, P., 1974. *Epilachna similis* Muls. (Coleoptera, Coccinellidae), a minor pest on maize in Ghana. Ghana Journal of Agricultural Science 7, 75–79.

Seshu Reddy, K.V., 1983. Sorghum stem borers in eastern Africa. Insect Science and Its Application 4, 3–10.

Setokuchi, O., 1975. The hibernation of *Longiunguis sacchari* on sorghums. Japanese Journal of Applied Entomology and Zoology 19, 296–297.

Shanthichandra, W., Gunasekera, S., Price, T., 1990. Diseases and pests of the winged bean (*Psophocarpus tetragonolobus*) in Sri Lanka. Tropical Pest Management 36 (4), 375–379.

Sharma, H.C., Lopez, V.F., 1990a. Biology and population dynamics of sorghum head bugs (Hemiptera: Miridae). Crop Protection 9 (3), 164–173.

Sharma, H.C., Vidyasagar, P., 1992. Orientation of males of sorghum midge, *Contarinia sorghicola* to sex pheromones from virgin females in the field. Entomologia Experimentalis et Applicata 64, 23–29.

Sharma, H.C., 1985. Screening for sorghum midge resistance and resistance mechanisms. In: Proceeding of the International Sorghum Entomology Workshop, Texas A&M University and ICRISAT, pp. 89–95.

Sharma, H.C., 2004. Insect-plant relationships of the sorghum midge, *Stenodiplosis sorghicola*. In: Bennett, J., Bentur, J.S., Pasalu, I.C., Krishnaiah, K. (Eds.), New Approaches to Gall Midge Resistance in Rice. Proceedings of the International Workshop, 22–24 November 1998. International Rice Research Institute and Indian Council of Agricultural Research, Hyderabad, India. Los Baños (Philippines). 195 p.

Sharma, H.C., Lopez, V.F., 1990b. Mechanisms of resistance in sorghum to head bug, *Calocoris angustatus*. Entomologia Experimentalis et Applicata 57 (3), 285–294.

Sharma, H.C., 1993. Host-plant resistance to insects in sorghum and its role in integrated pest management. Crop Protection 12, 11–34.

Sharma, H.C., Davies, J.C., 1988. Insect and Other Animal Pest of Millets. Patancheru, Andhra Pradesh 502 324, India: International Crops Research Institute for the Semi-arid Tropics. 86 pp.

Sharma, H.C., Leuschner, K., Vidyasagar, P., 1990. Factors influencing oviposition behaviour of the sorghum midge, *Contarinia sorghicola* Coq. Annals of Applied Biology 116, 431–439.

Sharma, H.C., Mukuru, S.Z., Hari Prasad, K.V., Manyasa, E., Pande, S., 1999. Identification of stable sources of resistance in sorghum to midge and their reaction to leaf diseases. Crop Protection 18, 29–37.

Sharma, H.C., Satyanarayana, M.V., Singh, S.D., Stenhouse, J.W., 2000. Inheritance of resistance to head bugs and its interaction with grain molds in *Sorghum bicolor*. Euphytica 112, 167–173.

Shufran, K.A., Burd, J.D., Anstead, J.A., Lushai, G., 2000. Mitochondrial DNA sequence divergence among greenbug (Homoptera: Aphididae) biotypes: evidence for host-adapted races. Insect Molecular Biology 9, 179–184.

Shufran, K.A., Peters, D.C., Webster, J.A., 1997. Generation of clonal diversity by sexual reproduction in the greenbug, *Schizaphis graminum*. Insect Molecular Biology 6, 203–209.

Singh, B.U., 1997. Screening for resistance to sorghum shoot bug and spider mites. In: Sharma, H.C., Singh, F., Nwanze, K.F. (Eds.), Plant Resistance to Insects in Sorghum. International Crops Research Institute for the Semi-arid Tropics (ICRISAT), Patancheru 502 324, Andhra Pradesh, India, pp. 52–59.

Singh, B.U., Rana, B.S., 1992. Stability of resistance to corn planthopper, *Peregrinus maidis* (Ashmead) in sorghum germplasm. Insect Science and Its Application 13, 251–263.

Singh, B.U., Seetharama, N., 2008. Host plant interactions of the corn planthopper, *Peregrinus maidis* Ashm. (Homoptera: Delphacidae) in maize and sorghum Agroecosystems. Arthropod-Plant Interactions 2, 163–196.

Singh, B.U., Padmaja, P.G., Seetharama, N., 2004. Biology and management of the sugarcane aphid, *Melanaphis sacchari* (Zehntner) (Homoptera: Aphididae), in sorghum: a review. Crop Protection 23, 739–755.

Singh, J., Mann, J.S., Dhaliwal, J.S., 1997. Biology of flower beetle *Chiloloba acuta* Wiedemann (Scarabaeidae: Coleoptera). Journal of Insect Science 102, 150–152.

Skidmore, P., 1985. The Biology of the Muscidae of the World. Kluwer Academic Publisher, 3300 AH Dordrecht, The Netherlands. 550 pp.

Srivastava, P.D., 1956. Studies on the copulatory and egg-laying habits of *Hieroglyphus nigrorepletus* Bol. (Orthoptera, Acrididae). Annals of the Entomological Society of America 49, 167–170.

Steck, G.J., Teetes, G.L., Maiga, S.D., 1989. Species composition and injury to sorghum by panicle feeding bugs in Niger. Insect Science and Its Application 10, 199–217.

Storey, H.H., 1938. Investigations of the mechanism of the transmission of plant viruses by insect vectors. II. The part played by puncture in transmission. Proceedings of the Royal Society of London B 125, 455–479.

Stretch, C., Kebreab, T., Edwards, S.B., 1980. The biology and control of the Welo bush-cricket, *Decticoides brevipennis* Ragge (Orthoptera: Tettigoniidae), a pest of cereals in Ethiopia. SINET 3 (1), 21–36.

Subbarayudu, B., Kalaisekar, A., Shyam Prasad, G., Bhagwat, V.R., 2010. Incidence of head bug, *Calocoris angustatus* on sorghum genotypes and the effect of abiotic factors on its population. Indian Journal of Plant Protection 38 (1), 22–24.

Sultana, M., Wagan, M.S., 2007. Life history and economic importance of *Hieroglyphus nigrorepleptus* (Acrididae: Orthoptera) from Pakistan. Journal of Entomology 4 (5), 379–386.

Sumner, L., Dorschner, K., Ryan, J., Eikenbary, R., Johnson, R., McNew, R., 1986. Reproduction of *Schizaphis graminum* (Homoptera: Aphididae) on resistant and susceptible wheat genotypes during simulated drought stress induced with polyethylene glycol. Environmental Entomology 15, 756–762.

Taley, Y.M., Thakare, K.R., 1980. Note on the population dynamics in carryover of *Chilo partelius* Swinhoe. Indian Journal of Agricultural Sciences 58, 635–637.

Tams, W.H.T., 1932. Newspecies of African Heterocera. Entomologist 65, 1241–1249.

Thomas, J.W., Mergendoller, J.R., 2000. Managing project-based learning: principles from the field. In: Paper Presented at the Annual Meeting of the American Educational Research Association, New Orleans.

Torikura, H., 1991. Revisional notes on Japanese Rhopalosiphum, with keys to species based on the morphs on the primary host. Japanese Journal of Entomology 59 (2), 257–273.

Tsai, J.H., 1996. Development and oviposition of *Peregrinus maidis* (Homoptera: Delphacidae) on various host plants. Florida Entomologist 79 (1), 19–26.

Tsai, J.H., Wilson, S.W., 1986. Biology of *Peregrinus maidis* with descriptions of immature stages (homoptera: Delphacidae). Annals of the Enlomological Society of America 79, 395–401.

Tsai, J.H., Perrier, J.L., 1993. Morphology of the digestive and reproductive systems of Peregrinus maidis (Homoptera: Delphacidae). Florida Entomologist 76 (3), 428–436.

Uchida, G.K., Mackey, B.E., Vargas, R.I., Beardsley, J.W., Hardy, D., Goff, M.L., Stark, J.D., 2006. Response of Nontarget Insects to Methyl Eugenol, Cue-lure, Trimedlure, and Protein Bait Bucket Traps on Kauai Island. (Hawaii, USA).

Unnithan, G.C., 1981. Aspects of sorghum shootfly reproduction. Insect Science and Its Application 2, 87–92.

Unnithan, G.C., Delobel, A.G.L., Raina, A.K., 1985. Off-season survival and seasonal carry-over of the sorghum shootfly, *Atherigona soeeata* Rondani (Diptera: muscidae) in Kenya. Tropical Pest Management 31, 115–119.

van Rensburg, N.J., 1973. Notes on the occurrence and biology of the sorghum aphid in South Africa. Journal of the Entomological Society of South Africa 36, 293–298.

Varma, A., Somadder, K., Kishore, R., 1978. Biology, bionomics and control of *Melanaphis indosacchari* David, a vector of sugarcane grassy shoot disease. Indian Journal of Agricultural Sciences 12, 65–72.

Vittum, P.J., Villani, M.G., Tashiro, H., 1999. Turfgrass Insects of the United States and Canada, second ed. Comstock, Ithaca.

Waladde, S.M., Kahoro, H.M., Ochieng, S.A., 1990. Sensory biology of *Chilo* spp. with specific reference to *C. partellus*. International Journal of Tropical Insect Science 11 (4–5), 593–602.

Walter, E.V., 1941. The biology and control of sorghum midge. Technical Bulletin 778, USDA, Washington D.C. pp. 27.

Webster, F.M., Phillips, W.J., 1912. The spring grain aphis or "greenbug". United States Bureau of Entomology Bulletin. 110, 1–153. https://archive.org/details/springgrainaphis110webs.

Weiland, A.A., Peairs, F.B., Randolph, T.L., Rudolph, J.B., Haley, S.D., Puterka, G.J., 2008. Biotypic diversity in Colorado Russian wheat aphid (Hemiptera: Aphididae) populations. Journal of Economic Entomology 101 (2), 569–574.

Weng, Y., Perumal, A., Burd, J.D., Rudd, J.C., 2010. Biotypic diversity in greenbug (Hemiptera: Aphididae): Microsatellite-based regional divergence and host-adapted differentiation. Journal of Economic Entomology 103 (4), 1454–1463.

Wijerathna, M.A.P., Edirisinghe, J.P., 1995. Preliminary observations on graminaceous aphids (Homoptera: Aphididae) of the Peradeniya University Park. Ceylon Journal of Science (Biological Sciences) 24 (1), 34–41.

Willott, E., Bew, L.K., Nagle, R.B., Wells, M.A., 1988. Sequential structural changes in the fat body of the tobacco hornworm, *Manduca sexta*, during the fifth larval stadium. Tissue Cell 20, 635–643.

Yadava, R.L., 1966. Oviparity in sugarcane aphid, *Longiunguis sacchari* Zehnt. (Aphidae: Homoptera). Current Science 1, 18.

Youm, O., Harris, K.M., Nwanze, K.F., 1996. *Coniesta ignefusalis* (Hompson), the millet stem borer: a handbook of information. Information Bulletin no. 46. Patancheru 502 324, Andhra Pradesh, India: International Crops Research Institute for the Semi-Arid Tropics. 60 pp.

Young, W.R., 1981. Fifty-five years of research on the sorghum shootfly. Insect Science and Its Application 2, 3–9.

Yue, Q., Johnson-Cicalese, J., Gianfagna, T.J., Meyer, W.A., 2000. Alkaloid production and chinch bug resistance in endophyte-inoculated chewings and strong creeping red fescues. Journal of Chemical Ecology 26, 279–292.

FURTHER READING

Brożek, J., 2006. Internal structure of the mouthparts in Coccinea (Hemiptera: Sternorrhyncha). Polish Journal of Entomology 75, 255–265.

INSECT–PLANT RELATIONSHIPS

HOST SIGNALING AND ORIENTATION

Phytophagous insects take the cue from an array of stimuli emanating from the plants to choose a suitable host plant (Zhang and Schlyter, 2004; Liu et al., 2011). Insects are equipped with sensory receptors enabling them to perceive these stimuli. Plant stimuli involved include, in varying proportions, visual, mechanical, gustatory and olfactory characteristics (Visser, 1986). Insects evolve sensory mechanisms to perceive specific cues from the right host plants and thus develop a specific host–plant relationship. Therefore the emergence of host specificity is the result of evolutionary changes in the insects' chemosensory systems (Jermy, 1984). The ovipositing female insects bear the responsibility of host plant recognition and selection to place the progeny larvae on a right host-plant. Visual and odor stimuli play an important role in host finding and oviposition process. Odors emanating from the plants may play a role in the orientation of its insect pests towards the plants and in ultimate recognition of the host plant for feeding and oviposition. In the process of odour recognition, the airborne volatile phytochemicals influence the interactions between plants and insect herbivores. Generally, the plants maintain a baseline level of volatile metabolites and are released from the surface of the leaf and other parts. Such baseline volatile phytochemicals that are normally emanating from the plants are called constitutive chemical reserves, which often include monoterpenes, sesquiterpenes, and aromatics, accumulate to high levels in specialized glands or trichomes (Paré and Tumlinson, 1997). When the leaves of the plant are mechanically damaged, odors consisting of a blend of saturated and unsaturated six-carbon alcohols, aldehydes, and esters are produced by autolytic oxidative breakdown of membrane lipids and are released.

While locating and recognizing host plant, the adult of sorghum shoot fly, *Atherigona soccata* utilizes specific blends of semiochemical cues for oviposition. Sorghum seedlings emit volatiles that are specific to both adult fly oviposition attraction and larval orientation/migration (Nwanze et al., 1998a,b). Younger susceptible seedlings are found to emit different quantities and blends of volatiles than resistant or older plants. Females of *A. soccata* are attracted both to the volatiles emitted by the susceptible seedlings, and to phototactic (optical) stimuli that may facilitate orientation to its host for oviposition (Nwanze et al., 1998a,b). Plant volatiles play an important role in determining susceptibility or resistance of a given sorghum cultivar (Padmaja et al., 2010). Identification of the specific volatiles can provide a better understanding of qualitative and quantitative differences in the volatile blends emanating from the resistant and susceptible sorghum plants. For example, DJ 6514, a susceptible cultivar to shoot fly in India and the shoot fly resistant cultivar IS 2312 emit different volatile profiles. IS 2312 emitted fewer volatiles than DJ 6514. The resistant cultivar IS 2312 did not emit α-pinene, (-)-(E)-caryophyllene, methyl salicylate, or octanal. Plant volatiles are collected using air entrainment chamber (Fig. 4.1). The volatile profiles of susceptible DJ 6514 and resistant IS 2312 are chromotographically

Insect Pests of Millets. http://dx.doi.org/10.1016/B978-0-12-804243-4.00004-5
Copyright © 2017 Elsevier Inc. All rights reserved.

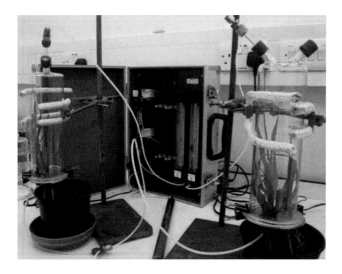

FIGURE 4.1

Air-entrainment chamber.

Courtesy of A. Kalaisekar.

recorded and presented in Fig. 4.2A and B, respectively. Female *A. soccata* produced a strong positive behavioral response when exposed to the volatiles collected from the susceptible cultivar DJ 6514 in an olfactometer bioassay.

Electrophysiologically active components in volatiles released by sorghum were detected by coupled gas chromatography–electroantennography, and the active peaks were identified by gas chromatography–mass spectrometry as (*Z*)-3-hexen-1-yl acetate, (−)-α-pinene, (−)-(*E*)-caryophyllene, methyl salicylate, octanal, decanal, 6-methyl-5-hepten-2-one, and nonanal (Figs. 4.2 and 4.3).

When an eight-component synthetic blend of the electroantennographic active compounds, at the same concentrations and ratios as in the natural headspace sample, was tested, *A. soccata* spent more time in the treated region of the olfactometer than in controls. Furthermore, when this synthetic blend and the natural headspace sample were tested in a choice test, the shoot flies did not show any preference for either of the two treatments, demonstrating that the synthetic blend had activity similar to the natural sample. This is the first identification of sorghum seedling volatiles functioning as kairomones in attracting *A. soccata* to hosts (Padmaja et al., 2010).

In case of spotted stem borer, *Chilo partellus,* electrophysiologically active components in volatiles released by sorghum are linalool, acetophenone, and 4-allylanisole (Birkett et al., 2006). The female moths could be using such cues to locate and recognize the sorghum plants as the host for oviposition and for subsequent larval survival.

In sorghum midge, *Stenodiplosis sorghicola* the females must identify the right stage of the flowering in sorghum to be able to oviposit and to sustain the specific feeding habit of the maggot that hatches from the egg. Under such circumstances, females utilize a combination of visual and chemical stimuli to locate sorghum crops. Generally, sorghum midge females prefer yellow followed by green, red, and blue colors (Sharma et al., 1988; Sharma and Franzmann, 2001a). Variation in spectral reflectance between sorghum genotypes may influence host selection by the sorghum midge. Sorghum midge response readily to host plant odor in combination with red and yellow colors. Therefore, both visual

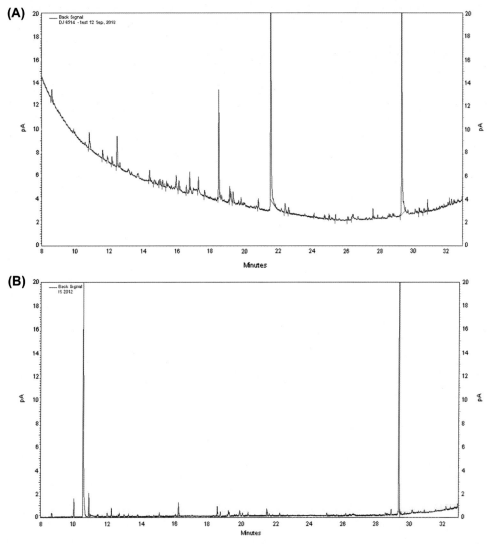

FIGURE 4.2

Volatile organic compounds identified in headspace samples of sorghum seedlings. (A) Sorghum cv. DJ 6514; (B) sorghum cv. IS 2312. Volatile compounds identified from the peaks are (−)-α-pinene, 6-methyl-5-hepten-2-one, octanal, (Z)-3-hexen-1-yl acetate, nonanal, methyl salicylate, decanal, (−)-(E)-caryophyllene.

Courtesy of P.G. Padmaja.

and chemical stimuli could influence host selection in sorghum midge (Sharma et al., 1990, 2002; Sharma and Franzmann, 2001a,b). Female midge flies receive chemical stimuli from viable pollen and receptive stigmata for oviposition. Such clues on the visual and chemical cues that elicit sharp responses in sorghum midge females could be exploited for developing suitable traps to monitor its abundance in combination with kairomones or pheromones (Sharma and Franzmann, 2001a,b; Sharma et al., 2002).

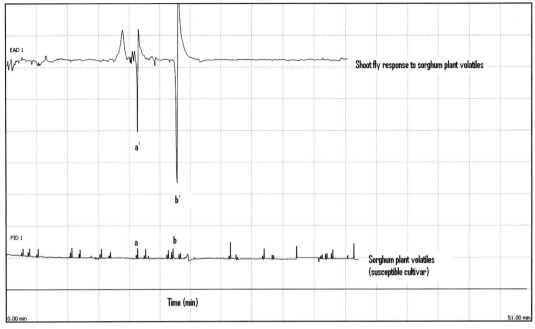

FIGURE 4.3

Electrophysiological response of female *Atherigona soccata* responses to sorghum cv. DJ 6514 seedling volatiles. The peaks marked as *a* and *b* are those volatile compounds that elicited corresponding responses *a'* and *b'* in electroantennogram coupled runs.

Courtesy of P.G. Padmaja.

HOST SPECIALIZATION

The host specialization in insects, especially sap feeders is driven by physiological efficiency, tolerance of fluctuating levels of host nitrogen, and enhanced success of dispersal or defense (Prestidge, 1982; Moran, 1986; Denno and Roderick, 1990). Insects use plant resources as shelter, food, and protection from predators. Use of host plant range is relatively different in the phytophagous insects. Plant-feeding insects are classified into two categories, generalists and specialists, according to the mode of host plant use by them. Generalist insects are those that use a wide range of plant species as their host, whereas specialist insects use a specified range of host plants. Phytophagous insects are again differentiated into three categories, monophagous, oligophagous, and polyphagous. The insect species which feed on single or several closely related plant species are termed as monophagous. The oligophagous species feed and develop on a wide range of plant species belonging to one or more closely related families. The polyphagous insects feed on wide range of plant species of unrelated plant families. The searching for and evaluation of suitable host plants is done mostly by female insects. Suitability for larval survival is the driving force for the host selection process. The decision-making process in insects for choosing various host plants or a single host plant is totally controlled by their nervous system. The chemical content of the plant greatly influences the host plant specialization in insects.

SHOOT FLIES

A vast array of feeding habits, such as saprophagy, coprophagy, carnivory, and phytophagy, is seen in the flies belonging to the Muscoidea. Saprophagy is the original ancestral feeding habit in muscid flies, and the other feeding habits arose later in various species groups (Sujatha et al., 2014). In Muscoidea, except for the family Fanniidae, other families, namely Muscidae, Anthomyiidae, and Scathophagidae, include species having phytophagous larval feeding habits (Sujatha et al., 2008). Notably, some of the species belonging mainly to the genera *Atherigona* (Muscidae) and *Delia* (Anthomyiidae) tend to oviposit on live plant hosts and the larvae feed on the decomposed plant tissue of that particular plant. For example, larvae of *Atherigona* spp., which are called shoot flies, survive on various grass species, and those of *Delia antiqua*, known as the onion fly; *Delia radicum*, called the cabbage root fly; and *Delia arambourgi* and *Delia flavibasis* feed inside the shoots of maize, wheat, pearl millet, tef, and some grasses. The larvae of several scathophagid species belonging to some genera such as *Delina*, *Hydromyza*, and *Neochirosia* are leaf miners of many plant species. The leaf miners can be truly phytophagous by virtue of their feeding on live plant tissues. But to consider the feeding habits especially of the species of *Atherigona* as true phytophagy is arguable and warrants great depth of understanding. This is because the female fly oviposits only on live plants, whereas the larvae sever the inner base of the growing shoot to have decomposed plant tissue for feeding. Female flies of various species of *Atherigona* show remarkable host selection and specialization for ovipositioning.

Species of *Atherigona* are oligophagy and are known to develop on grasses (Poaceae) and sedges (Cyperaceae). Cultivated sorghum, *Sorghum bicolor* is the most preferred host of sorghum shoot fly, *A. soccata* (Davies and Seshu Reddy, 1981). Apart from sorghum, *A. soccata* is found to survive on about 22 plant species belonging to both grasses and sedges. The host plants include, cultivated grasses namely *Zea mays*, *Pennisetum typhoides*, *Setaria italica*, *Panicum* spp., *Echinochloa frumentacea* and several wild grasses like, *Echinochloa colonum*, *Eriochloa procera*, *Cymbopogon* sp. and *Paspalum* spp., and species of *Brachiaria*, *Cynodon*, *Echinochloa*, *Eragrostis*, *Panicum*, *Pennisetum*, *Setaria* and other *Sorghum* spp. (Seshu Reddy and Davies, 1977; Davies and Seshu Reddy, 1981; Gahukar, 1991). Several wild graminaceous plants have been reported as hosts of the sorghum shoot fly in various parts of Africa. *Sorghum halepense* is by far the most important alternative host plant other than the cultivated sorghum (Davies and Seshu Reddy, 1981). Three grasses, *Digitaria ascendens*, *Brachiaria reptans* and *Eleusine indica* are the possible alternative food-plants of *A. soccata* in Thailand. The wide range of wild grass hosts emphasizes the adaptability and potential danger of this insect (CABI, 2007). Small populations persist on wild food-plants in the absence of sorghum.

Other species of *Atherigona* are also recorded on several grasses and sedges (see Chapter 2) but the host range is comparatively smaller than *A. soccata*. The possibility of evolution of feeding mechanisms especially in terms of variations in cephalopharyngeal structures of maggots, in host expansion needed to be investigated. *A. approximata*, infests pearl millet, *A. falcata*, *A. naqvii*, infests wheat, maize and barley in India and Pakistan, *P. glacum* in Nigeria, wheat in Saudi Arabia. *A. oryzae*, which is a widespread pest of rice in Pakistan. *A. exigua* and *A. varia* are rarely noticed species. *A. orientalis* is also found in sorghum and maize, but is only saprophagous (Pont, 1972). *A. eriochloae* is able to survive in ratooned sorghum (Davies et al., 1980). In addition to *A. soccata*, males of *A. campestris*, *A. gilvifolia*, *A. tomentigera*, *A. secrecauda* collected from sorghum in Uganda (Baliddawa and Lyon, 1974).

STEM BORERS

The spotted stemborer *Chilo partellus* is an oligophagy and feeds on species of grasses and sedges which include several important cultivated cereals, especially maize, sorghum, rice, sugarcane, and millets. It survives on several grasses including sudan grass (*Sorghum vulgare sudanense*), napier grass (*Pennisetum purpureum*), and *Sorghum arundinaceum*. There are about 19 cultivated and wild plants identified as host plants of the stem borers (Kfir, 1997, 2002; Khan et al., 2000; Kfir et al., 2002; Assefa et al., 2006; Swami and Bajpai, 2006; Matama-Kauma et al., 2008; Bhagwat et al., 2011; Shyam Prasad et al., 2011, 2015). Among the cultivated crops maize, sorghum, sugarcane and barley are the most preferred hosts and best for the development of *C. partellus* (Swami and Bajpai, 2006; Bhagwat et al., 2011; Subbarayudu et al., 2011; Shyam Prasad et al., 2010, 2011, 2015). Larva undergoes diapause in the cooler regions where winter temperatures fall below 10°C. In the warmer regions, advancement of generations takes place in several wild grasses.

The stem borers such as *Busseola* sp., *Chilo partellus*, *Sesamia calamistis* and *Eldana saccharina* take shelter in indigenous sedges present in irrigation canals for off-season survival. *Busseola* sp. was the predominant stem borer of sugarcane, in parts of Africa (Assefa et al., 2006). This species is a common stem borer that attack maize and sorghum in eastern, northern, southern and western Ethiopia (Getu et al., 2001). *C. partellus* is observed to oviposit on non host plants such as cowpea and cassava when intercropped with sorghum. The eggs were able to hatch on cowpea but the number of neonate larvae that arrived on the sorghum host diminished with distance. *C. partellus* oviposition on non-host crops is one of the mechanisms for its reduced abundance in intercropping (Ampong-Nyarko et al., 1994). In maize ecosystem, larvae of *C. partellus* distributed in aggregated manner due to environmental heterogeneity (Jalali and Singh, 2001).

Cultivated host plants showed higher incidences of whorl and stem damage than in wild grasses such as *Hyparrhenia tamba* and, *Panicum maximum* (Rebe et al., 2004). Similarly, the other species of stem borers such as *Sesamia calamistis, S. inferens, Chilo orichalcociliellus, C. sacchariphagus, C. auricilius* and *C. infuscatellus* are also survive on several grasses. In many parts of Africa, the invasive *C. partellus* displaced the native *C. orichalcociliellus, Busseola fusca* and *Sesamia inferens* (Kfir et al., 2002). On a daily basis, *C. partellus* consumed more maize and sorghum than *C. orichalcociliellus*. A few *C. orichalcociliellus* survived to the pupal stage in napier and guinea grasses, whereas no *C. partellus* survived. The shorter developmental period of *C. partellus* may give this species a competitive advantage over the slower developing *C. orichalcociliellus*. However, the ability of *C. orichalcociliellus* to complete development in two native grasses in which *C. partellus* did not survive may provide a refuge that has allowed *C. orichalcociliellus* to escape extirpation from the coastal area of Kenya (Ofomata et al., 2000).

SHOOT BUG

Peregrinus maidis is an exception of being oligophagy in an otherwise monophagous species group of delphacid planthoppers (Denno and Roderick, 1990). On the contrary, the cicadellid leafhoppers are largely of oligophagy (Whitcomb et al., 1987). The planthoppers owe the host specificity to their ability to synchronize the life-histories to the fluctuating nitrogen levels of host plant (Moran, 1986). Sorghum is the ancestral host of shoot bug, *Peregrinus maidis* (Nault, 1983a,b). *P. maidis* is one of the most extensively studied plant hoppers chiefly due to its association with MStrV and MMV as vector (refer to section "Morphology and Anatomy" in Chapter 3).

P. maidis principally prefers to colonize on sorghum and maize (Tsai, 1975; Nault, 1983a,b). It transmits MStrV and MMV to sorghum and maize and also to Itch grass (*Rottboellia exaltata*) (Tsai, 1975; Nault and Knoke, 1981; Tsai and Falk, 1988). *P. maidis* is known to survive on a vast array of species spectrum in the plant families such as Cannaceae, Commelinidae, Cyperaceae, Gramineae and Sterculiaceae (Singh and Seetharama, 2008). Many of the plant species of those families serve as spring, summer or overwintering, accidental, or alternate hosts for oviposition and feeding by *P. maidis* (Tsai, 1996; Singh and Seetharama, 2008). For example, *P. maidis* utilizes *Pennisetum typhoides*, *Sorghum halepense*, *Setaria italica*, *Echinochloa frumentacea* and *Paspalum scrobiculatum* as breeding or feeding hosts (Chelliah and Basheer, 1965; Namba and Higa, 1971). The three most preffered hosts of *P. maidis* are sorghum, maize and itchgrass (Catindig, 1993; Catindig et al., 1996). In the times of dearth for preferred hosts, *P. maidis* oviposits on other related plant species such as Cox's tear, Johnsongrass, and sugarcane (Napompeth, 1973). Total number of eggs produced is considerably greater on sorghum (Tsai, 1996). Adults are found to survive on rice, rye, oats and sugarcane while nymphs fail to develop. Survival rate of neonate-to-adult is found to be high on sorghum, maize, itchgrass (Catindig, 1993; Catindig et al., 1995; Tsai, 1996).

APHIDS

At present, nearly equal proportions of aphids species have colonized woody and herbaceous plants (Fig. 4.4). The largest numbers of aphid species are recorded on Apiaceae, Astraceae, Rosaceae, and Coniferae (Fig. 4.5). The aphids that live only on herbaceous plants are (Blackman and Eastop, 2006):

1. Saltusaphidinae on Cyperaceae and Juncaceae,
2. Siphini on Poaceae, and
3. Tramini on roots of Asteraceae.

Melanaphis sacchari

The sugarcane aphid, *M. sacchari* is an oligophagy and largely restricted to the species of the *Saccharum*, *Sorghum*, *Oryza*, *Panicum*, and *Pennisetum* (Denmark, 1988). *M. sacchari* prefers to colonize on sugarcane as its principal host plant. It reportedly entered Central America around 1932 and spread to south, reported from Argentina and Uruguay (Blackman and Eastop, 2006). It is a regular pest on sorghum in South Africa (Muller and Scholl, 1958; Matthee and Oberholzer, 1958; Matthee, 1962; Blackman and Eastop, 2006). There are several host plants that are known to support *M. sacchari* such as *Cynodon dactylon*, *Miscanthus chinensis*, *Setaria italic*, *Sorghum halepense*, *Sorghum verticilliflorum* and *Paspalum sanguinale* (Wilbrink, 1922; Setokuchi, 1973; van Rensburg, 1973; Agarwala, 1985; van Rensburg and van Hamburg, 1975; Kawada, 1995). However, sorghum and Johnsongrass, *Sorghum halepense* are highly suitable host plants apart from sugar cane (Armstrong et al., 2015). *M. sacchari* is anholocyclic throughout most of its range of occurrence, but for presence of some monoecious holocyclic forms on sorghum and sugarcane in India (Yadava, 1966; David and Sandhu, 1976).

GREENBUG

The greenbugs need special mention here in the context of host plant association, chiefly because *Schizaphis graminum* is exclusively a grass feeder, even during the holocycly (refer to "Reproductive Behavior and Life History" section, Chapter 3). There are more than 80 grass species, including several cultivated cereals, millets, and turfgrasses, that support the survival of *S. graminum*. *H. vulgare*,

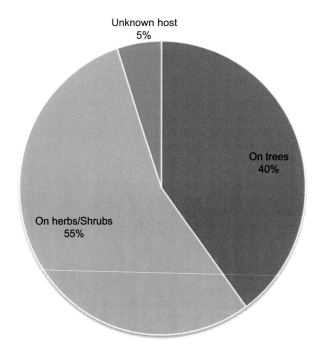

FIGURE 4.4

Proportion of aphid species living on various types of plants.

Data from Blackman, R.L., Eastop, V.F., 2006. Aphids on the World's Herbaceous Plants and Shrubs. Wiley and Sons, Hoboken, USA. Courtesy of A. Kalaisekar.

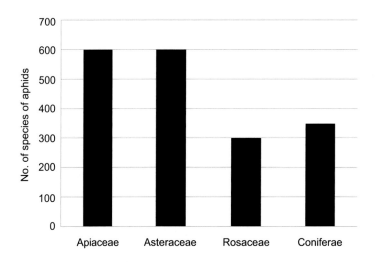

FIGURE 4.5

Largest numbers of species of aphids recorded on certain plant taxonomic groups.

Data from Blackman, R.L., Eastop, V.F., 2006. Aphids on the World's Herbaceous Plants and Shrubs. Wiley and Sons, Hoboken, USA. Courtesy of A. Kalaisekar.

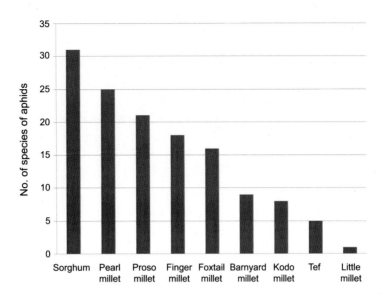

FIGURE 4.6

Numbers of species recorded on various millet crops.

Courtesy of A. Kalaisekar.

Z. mays, A. sativa, O. sativa, S. cereale, S. bicolor, Triticum aestivum, Poa pratensis, and *Paspalum vaginatum* are some of the cultivated cereals and millets. The list of hosts also includes wild grasses such as *Agropyron* spp., *Agrostis* spp., *Alopecurus* spp., *Andropogon* spp., *Bothrichloa* spp., *Bouteloua* spp., *Bromus* spp., *Chloris* spp., *Cynodon* spp., *Dactylis* spp., *Digitaria* spp., *Echinochloa* spp., *Eleusine* spp., *Elymus* spp., *Eragrostis* spp., *Festuca* spp., *Lolium* spp., *Panicum* spp., *Pennisetum* spp., *Phleum* spp., *Setaria* spp., *Sorghastrum* spp., *Sorghum* spp., *Stipa* spp., *Tripsacum* spp., and *Zea* spp. (Webster and Phillips, 1912; Dahms et al., 1936; Michels, 1986; Nuessly and Nagata, 2005; Blackman and Eastop, 2006; Weng et al., 2010; Royer et al., 2015).

The greenbugs are known to exist in several biotypic forms on wheat, sorghum, and turfgrasses (Porter et al., 1997; Nuessly et al., 2008; Royer et al., 2015). The biotypes recognized in the United States are A, B, E, G, H, and J on wheat; biotypes C, D, I, and K on sorghum; biotype F on Canada bluegrass, and biotype "Florida isolate" on seashore paspalum (Royer et al., 2015).

There are several species of aphids recorded on millets across the world (Fig. 4.6) and the following is a list of aphid species and their associated millet host plants (Blackman and Eastop, 2006).

1. *Anoecia cornicola*: pearl millet
2. *Anoecia corni*: barnyard millet, finger millet, foxtail millet, pearl millet, sorghum, proso millet
3. *Anoecia fulviabdominalis*: pearl millet, sorghum, barnyard millet
4. *Anoecia krizusi*: sorghum
5. *Anoecia vagans*: pearl millet, proso millet
6. *Aphis fabae*: sorghum
7. *Aphis gossypii*: finger millet, pearl millet, sorghum, kodo millet
8. *Aphis spiraecola*: finger millet
9. *Brachycaudus helichrysi*: finger millet

10. *Diuraphis noxia*: sorghum
11. *Forda hirsuta*: pearl millet, sorghum
12. *Forda marginata*: proso millet
13. *Forda orientalis*: pearl millet, sorghum
14. *Geoica lucifuga*: finger millet, sorghum, kodo millet
15. *Geoica utricularia*: pearl millet, proso millet
16. *Hysteroneura setariae*: finger millet, little millet, pearl millet, sorghum, kodo millet, tef
17. *Melanaphis pyraria*: pearl millet
18. *M. sacchari*: foxtail millet, pearl millet, sorghum, proso millet
19. *Melanaphis sorghi*: finger millet, sorghum
20. *Myzus persicae*: pearl millet, sorghum
21. *Neomyzus circumflexus*: pearl millet
22. *Paracletus cimiciformis* subsp. *panicumi*: proso millet
23. *Paracletus cimiciformis*: foxtail millet, pearl millet, sorghum
24. *Protaphis middletonii*: pearl millet
25. *Rhopalosiphum maidis*: barnyard millet, finger millet, foxtail millet, pearl millet, sorghum, proso millet, kodo millet
26. *Rhopalosiphum nymphaeae*: pearl millet
27. *Rhopalosiphum padi*: barnyard millet, finger millet, foxtail millet, pearl millet, sorghum, proso millet
28. *Rhopalosiphum rufiabdominale*: barnyard millet, foxtail millet, pearl millet
29. *S. graminum*: finger millet, foxtail millet, pearl millet, sorghum, proso millet, tef
30. *Sipha elegans*: foxtail millet, pearl millet
31. *Sipha flava*: barnyard millet, foxtail millet, sorghum
32. *Sipha maydis*: finger millet, pearl millet, sorghum, proso millet
33. *Sitobion africanum*: sorghum
34. *Sitobion avenae*: finger millet, foxtail millet, pearl millet, sorghum, proso millet
35. *Sitobion graminis*: kodo millet
36. *Sitobion indicum*: kodo millet
37. *Sitobion leelamaniae*: finger millet, pearl millet, sorghum
38. *Sitobion miscanthi*: finger millet, foxtail millet, pearl millet, sorghum
39. *Sitobion pauliani*: pearl millet
40. *Smynthurodes betae*: tef
41. *Tetraneura nigriabdominalis*: foxtail millet, sorghum, finger millet, barnyard millet
42. *Tetraneura nigriabdominalis* subsp. *shanxiensis*: sorghum
43. *Tetraneura triangula*: sorghum
44. *Tetraneura africana*: pearl millet
45. *Tetraneura basui*: finger millet, pearl millet, kodo millet
46. *Tetraneura caerulescens*: foxtail millet, proso millet
47. *Tetraneura capitata*: sorghum
48. *Tetraneura chui*: sorghum
49. *Tetraneura fusiformis*: barnyard millet, finger millet, foxtail millet, pearl millet, proso millet, sorghum, kodo millet
50. *Tetraneura javensis*: finger millet, foxtail millet, proso millet, sorghum
51. *Tetraneura ulmi*: proso millet, sorghum
52. *Tetraneura yezoensis*: barnyard millet, finger millet, foxtail millet, pearl millet, proso millet.

MIDGES

The sorghum midge, *Stenodiplosis sorghicola*, survives only on the members of the genus *Sorghum* (Harris, 1979; Franzmann and Hardy, 1996). These include Johnson grass (*S. halepense*), Columbus grass (*Sorghum almum*), and Sudan grass (*Sorghum sudanense*). Johnson grass has been recognized as important in the population dynamics of the sorghum midge and acts as an early season host for midge (Harris, 1961; Teetes, 1985; Franzmann et al., 2006). The peak adult emergence of overwintering midge coincides with the widespread flowering of Johnson grass in an area after spring rains. Johnson grass acts as a host for two or three generations, and then the midge migrates to infest flowering grain-sorghum panicles. Flowering grain-sorghum crops are the most susceptible to midges for egg-laying at the time of anthesis.

Females do not lay eggs in the spikelets of *Sorghum amplum*, *Sorghum bulbosum*, and *Sorghum angustum* under no-choice conditions (Franzmann and Hardy, 1996; Sharma and Franzmann, 2001a,b). Panicles of *S. halepense* are more attractive to the females than those of *Sorghum stipoideum*, *Sorghum brachypodum*, *S. angustum*, *Sorghum macrospermum*, *Sorghum nitidium*, *Sorghum laxiflorum*, and *S. amplum*. Sorghum panicles at the half-anthesis stage are more attractive to females than those at the pre- or postanthesis stage (Sharma et al., 1990).

PEST–HOST EVOLUTIONARY RELATIONSHIP
SHOOT FLIES

Saprophagy is the ancestral feeding habit in muscid flies (Sujatha et al., 2014). The majority of muscid species are still practicing saprophagy. There have been several twists and turns that lead to other modes of feeding such as coprophagy, carnivory, hematophagy, and phytophagy in muscids. Currently, there are four muscoid families, namely Anthomyiidae, Fanniidae, Muscidae, and Scathophagidae (Sujatha et al., 2008). Among these families, except Fanniidae, which comprises all saprophagous species, Anthomyiidae, Muscidae, and Scathophagidae have some of the species that are phytophagous. There are some true phytophagous species in Anthomyiidae and Scathophagidae. For example, in Anthomyiidae, several species are leaf miners; notably *Delia echinata* (Seguy, 1923), *Delia florilega* (Zetterstedt, 1845), and *Pegomya* spp. are leaf miners on plants belonging to Amaranthaceae, Brassicaceae, Caryophyllaceae, Chenopodiaceae, Polemoniaceae, Primulaceae, and Solanaceae (Pitkin et al., 2007). In the Scathophagidae, there are species such as *Delina nigrita* (Fallén, 1819), a leaf miner on orchids; *Hydromyza livens* (Fabricius, 1794) and *Norellia spinipes* (Meigen, 1826), leaf miners on water lily (Robbins, 1984); and *Parallelomma* species, leaf miners on liliaceous and orchidaceous plants (Pitkin et al., 2007). In the Muscidae, the flies belonging to the genus *Atherigona* are phytophagous. Unlike the true phytophagy in the species of Anthomyiidae and Scathophagidae that feed on live plant tissues, the larvae of *Atherigona* species feed on decomposed plant tissue inside the plant shoot. The evolution of phytophagy in the Muscidae is not well understood. An analysis of the varying degrees of feeding habits found in the four subfamilies of Muscidae is important to understand the evolution of phytophagy in Muscidae. There are some leads in this aspect with respect to feeding habits in the Scathophagidae (Sujatha et al., 2014). The evolution of herbivory in Drosophilidae involves major behavioral changes in chemosensory systems (Benjamin et al., 2014).

There are again differences between the two subgenera of *Atherigona* with reference to larval feeding habits. Larvae of the subgenus *Atherigona* s.str. cause the plant tissue to decompose, as a result of severing the innermost growing shoot whorl of the host plant, and feed upon it. All the species of this subgenus prefer to feed on only the graminaceous plants. The larvae of another subgenus, *Acritochaeta*, usually feed on decomposed plant tissues of fruits such as tomato, pepper, and chilies that are dicot plants. Therefore, the feeding habits of the species of *Atherigona* are certainly a matter of great ambiguity. The pest–host coevolutionary aspects of *Atherigona* and its host plants are poorly understood and need to be thoroughly investigated. The feeding habits in muscid flies as a whole are associated with a reduction in the number of larval instars (Skidmore, 1985).

STEM BORERS

In general, the concealed feeding habits, such as mining, boring, and soil root feeding, in lepidopteran larvae developed before leaf feeding. This is corroborated by the fact that the more advanced families in the phylogenetic arrangements show leaf feeding in Lepidoptera. The stem boring by lepidopterans in Gramineae is a specialized feeding habit and might have evolved during the early periods of their association in the Mesozoic period (Gould and Shaw, 1983). The millets, including sorghum, are the indispensable hosts of a number of lepidopteran borers (as given in Chapter 2).

APHIDS

Evolutionarily older tree groups, which were on land before the appearance of herbaceous plants, are the most preferred hosts of aphids (Blackman and Eastop, 1984a,b, 1990, 1994a,b, 2000, 2006, 2007a,b).

Aphids are predominantly a northern temperate group, with remarkably few species in the tropics (Blackman and Eastop, 2006). Cyclical parthenogenesis is proposed as the reason for the aphids not being able to diversify into the tropics (Blackman and Eastop, 1984a,b, 1990, 2000, 2006). Cyclical parthenogenesis is an adaptive trait in aphids to exploit the short-lived growth flushes of temperate plants (Blackman and Eastop, 1990, 1994a,b, 2000, 2006). The great diversity of forest plants in tropics negates the successful establishment of short-lived and host-specific aphids (Dixon et al., 1987). Aphids are absent from several tropical plants, notably all the known plants of Dipterocarpaceae, *Swietenia mahogoni* (Meliaceae), and *Dalbergia nigra* (Fabaceae) (Blackman and Eastop, 2006). Aphids lose the sexual reproductive phase in the tropics and thereby fail to diversify into different species in tropical regions. The tropical belt around the globe acts as a barrier for the aphid species to establish in the southern temperate regions (Blackman and Eastop, 1990, 1994a,b, 2000, 2006, 2007a,b).

In the greenbug, *S. graminum*, the molecular data of biotypes in North America reveals a different divergence and thus points to the fact that the establishment of host relationships of *S. graminum* was rooted much before the start of modern agriculture (Shufran et al., 1997, 2000; Blackman and Eastop, 2006).

SHOOT BUG

Peregrinus has an African evolutionary origin and is more closely linked to *Sorghum* than *Zea* (Nault, 1983a,b). Sorghum is considered to be the ancestral original host plant of the shoot bug, *P. maidis* (Nault, 1983a,b). *P. maidis* appears to tolerate only tropical climates and is well adapted to survive on sorghum, a tropical plant species (Nault, 1983a,b). The bugs are recorded on other tropical grasses such as *S. italica*, *E. colonum*, *P. scrobiculatum*, and *P. typhoides*, in India (Chelliah and Basheer, 1965), and *R. exaltata* in Florida, USA (Nault, 1983a,b).

BIOGEOGRAPHY OF INSECTS AND HOST PLANTS

Biogeography deals with the distribution of species and ecosystems in geographic space and geological time. The two main branches of biogeography are phytogeography and zoogeography. In this section, the distribution and spread of insect pests that are specific to millets are discussed based on the zoogeographic realms of the world (Fig. 4.7).

Zoogeography is the science of the study of the distribution of animal life.

The following are the six major zoogeographic regions of the world.

1. **Palearctic region**: Europe, north temperate Asia, and north to the border of Sahelian Africa
2. **Ethiopian or Palaeotropical region**: all tropical and South Africa, Madagascar, and the Mascarene Islands
3. **Oriental region**: all Asia south of the Palearctic limits, India, and southeast Asia
4. **Australian region**: Australia, New Zealand, Papuan Islands, and the islands of Oceanica
5. **Nearctic region**: temperate and arctic North America and Greenland
6. **Neotropical region**: Central America, West Indian Islands, and South America.

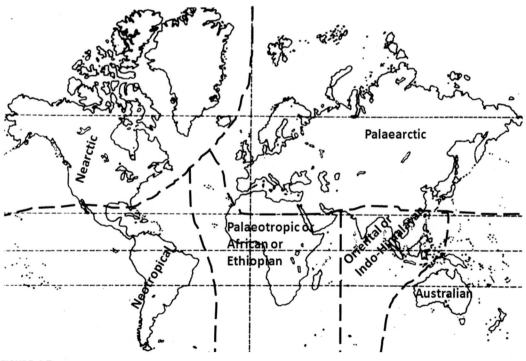

FIGURE 4.7

Zoogeographic regions of the world.

Courtesy of A. Kalaisekar.

When we discuss the zoogeography of the insects, we restrict it to the insect pests that are specific to millet feeding. The specific insects are mainly shoot flies, stem borers, shoot bug, and midges.

SHOOT FLIES

The distribution and spread of shoot flies of the genus *Atherigona* is largely restricted to the tropical areas covering the whole of the Palaeotropic and Oriental regions and parts of the Palearctic region. Though there are records of the occurrence of *Atherigona* species such as *Atherigona reversura* in the Nearctic region, the species associated with cultivated millets are completely absent from other zoogeographic regions. Almost all the millets cultivated are of either African or Asian origin. The spread of millets to the other parts of the world has not helped the establishment of the shoot flies in those new areas.

STEM BORERS

C. partellus is an endemic pest of India and invaded Africa around or after 1930. Currently this species is established in much of eastern and southern Africa. *C. partellus* partially displaced a native species, *C. orichalcociliellus*, in some areas of Africa. It coexists with another stem borer, *B. fusca*, in some areas and partially displaced it in parts of Africa (Polaszek, 1998; Kfir, 1997). The species composition of *B. fusca*, *C. partellus*, and *S. calamistis* differs from place to place (Okech et al., 1994). *C. partellus* usually prefers lower altitudes, whereas *B. fusca* occurs at higher altitudes.

Chilo auricilius, the gold-fringed borer, is another important borer occurring in the Oriental region. *Chilo polychrysus* (Meyrick) is a species superficially similar to *C. auricilius*, and their distribution ranges overlap in India, Indonesia, and Thailand.

Maliarpha separatella is widely distributed in Africa and Asia.

Sesamia inferens is a widely distributed species in the Oriental region and east to the Palearctic region.

SHOOT BUG

This species is widely distributed in the tropics and coastal areas of the subtropical and temperate regions of all continents. Tropical Africa is the genetic headquarters of *Peregrinus* and it spread into India and southeast Asia in prehistoric times and later entered the New World (Nault, 1983a,b).

MIDGES

The sorghum midge, *S. sorghicola*, is native to Africa and spread to India. Currently, the pest is distributed worldwide. During the second half of the 19th century, *S. sorghicola* was a serious pest of sorghum cultivation in the United States. The pest might have spread to America along with the introduction of sorghum from Africa (Walter, 1941). As early as 1890s the midge was so serious that it wiped out sorghum in many southern states of the United States (Hansen, 1923; Walter, 1941).

The other midge that attack millets is *Geiromiya penniseti* and it is distributed in India.

APHIDS

Aphids are predominantly a northern temperate group, with remarkably few species in the tropics (Blackman and Eastop, 1984a, 1994a, 2006). Cyclical parthenogenesis is proposed as the reason for aphids not being able to diversify into the tropics (Blackman and Eastop, 2000, 2006). Aphids lose the sexual reproductive phase in the tropics and thereby fail to diversify into different species in tropical regions. The tropical belt around the globe acts as a barrier for the aphid species to establish in the southern temperate regions (Blackman and Eastop, 2000, 2006).

S. graminum is of Palearctic origin. It is widely distributed in southern Europe (no records of presence in northern Europe), the Middle East, Central Asia, Africa, India, Nepal, Pakistan, Thailand, Korea, Taiwan, Japan, and the Americas (Blackman and Eastop, 2006).

REFERENCES

Agarwala, B.K., 1985. Notes on some aphids (Homoptera: Aphididae) affecting economically important plants on Bhutan. Indian Agriculture 27, 261–262.

Ampong-Nyarko, K., Seshu Reddy, K.V., Nyang'or, R.A., Saxena, K.N., 1994. Reduction of insect pest attack on sorghum and cowpea by intercropping. Entomologia Experimentalist et Applicata 70, 179–184.

Armstrong, J.S., Rooney, W.L., Peterson, G.C., Villenueva, R.T., Brewer, M.J., Sekula-Ortiz, D., 2015. Sugarcane Aphid (Hemiptera: Aphididae): Host Range and Sorghum Resistance Including Cross-Resistance From Greenbug Sources. Journal of Economic Entomology 108 (2), 576–582.

Assefa, Y., Conlong, D.E., Mitchell, A., 2006. First records of the stem borer complex (Lepidoptera: Noctuidae; Crambidae; Pyralidae) in commercial sugarcane estates of Ethiopia, their host plants and natural enemies. In: Proceedings of the 80th Annual Congress of the South African Sugar Technologists' Association, Durban, South Africa, 18–20 July 2006, pp. 202–213.

Baliddawa, C.W., Lyon, W.F., 1974. Sorghum shoot fly species and their control in Uganda. PANS 20 (1), 20–22.

Benjamin, G.H., Mitchell, R.F., Lapoint, R.T., Faucher, C., Hildebrand, J.G., Whiteman, N.K., 2014. Evolution of herbivory in Drosophilidae linked to loss of behaviors, antennal responses, odorant receptors, and ancestral diet. PNAS 112 (10), 3026–3031.

Bhagwat, V.R., Shyam Prasad, G., Kalaisekar, A., Subbarayudu, B., Hussain, T., Upadhyaya, S.N., Daware, D.G., Rote, R.G., Rajaram, V., 2011. Evaluation of some local sorghum checks to shoot fly, *Atherigona soccata* Rondani and Stem borer, *Chilo partellus* Swinhoe resistance. Annals of Arid Zone 50 (1), 47–52.

Birkett, M.A., Chamberlain, K., Khan, Z.R., Pickett, J.A., Toshova, T., Wadhams, L.J., Woodcock, C.M., 2006. Electrophysiological responses of the lepidopterous stemborers *Chilo partellus* and *Busseola fusca* to volatiles from wild and cultivated host plants. Journal of Chemical Ecology 32 (11), 2475–2487.

Blackman, R.L., Eastop, V.F., 1984a. Aphids on the World's Crop an Identification Guide. Wiley British Museum London, United Kingdom.

Blackman, R.L., Eastop, V.F., 1984b. Aphids on the world's Crops. Wiley, Chichester & New York, 466 pp.

Blackman, R.L., Eastop, V.F., 1990. Biology and taxonomy of the aphids transmitting barley yellow dwarf virus. III. Ecology and epidemiology. In: Burnett, P.A. (Ed.), World Perspective on Barley Yellow Dwarf. CYMMYT, Mexico, D.F., Mexico, pp. 197–214.

Blackman, R.L., Eastop, V.F., 1994a. Aphid on the World's Trees an Identification and Information Guide. CAB International Wallingford, United Kingdom.

Blackman, R.L., Eastop, V.F., 1994b. Aphids on the World's Trees. CAB International, Wallingford, 987 pp.

Blackman, R.L., Eastop, V.F., 2000. Aphids on the World's Crops, second ed. Wiley, Chichester, 466 pp.

Blackman, R., Eastop, V.F., 2006. Aphids on the World's Herbaceous Plants and Shrubs. Wiley and Sons, Hoboken, USA.

Blackman, R.L., Eastop, V.F., 2007a. Taxonomic issues. In: van Emden, V.F., Harrington, R. (Eds.), Aphids as Crop Pests. CAB International, Wallingford, UK, pp. 1–29.

Blackman, R.L., Eastop, V.F., 2007b. Taxonomic issues. In: van Emden, H.F., Harrington, R. (Eds.). Aphids as Crop Pests, CABI, UK, pp. 1–29.

CABI, 2007. Crop Protection Compendium. Online http://www.cabicompendium.org/CABI,UK.

Catindig, J.D., Barrion, A.T., Litsinger, J.A., 1995. Evaluation of rice, maize, and 56 rice field weeds of planthopper, *Peregrinus maidis* (Ashmead). International Rice Research Newsletter 20, 25–26.

Catindig, J.D., Barrion, A.T., Litsinger, J.A., 1996. Plant host range and life history of the corn delphacid, *Peregrinus maidis* (Ashmead) (Hemiptera: Delphacidae). Asia Life Sciences (Philippinnes) 5, 35–46.

Catindig, J.L., 1993. Use of Taxonomic Affinities of Plants and Insects to Predict the Host Range of Six Selected Oligophagous Herbivorous Pests of Rice (M.Sc. thesis). University of Philippines, Los Banos, Laguna, Philippines. 176 pp.

Chelliah, S., Basheer, M., 1965. Biological studies of *Peregrinus maidis* (Ashmead) (Araeopidae: Homoptera) on sorghum. Indian Journal of Entomology 27, 466–471.

Cugala, D., Omwega, C.O., 2001. Cereal stemborer distribution and abundance, and introduction and establishment of *Cotesia flavipes* Cameron (Hymenoptera: Braconidae) in Mozambique. Insect Science and Its Application 21 (4), 281–287. (Status and advances in biological control of cereal stemborers in Africa.); 26 ref.

Dahms, R.G., Snelling, R.O., Fenton, F.A., 1936. Effect of different varieties of sorghum on biology of the chinch bug. Journal of the American Society of Agronomy 28, 160–161.

David, S.K., Sandhu, G.S., 1976. New oviparous morph on *Melanaphis sacchari* (Zehntner) on sorghum. Entomologist's Record 88, 28–29.

Davies, J.C., Seshu Reddy, K.V., 1981. Shoot fly species and their graminaceous hosts in Andhra Pradesh, India. Insect Science and its Application 2, 33–37.

Davies, J.C., Reddy, K.V.S., Skinner, J.D., 1980. Attractants for Atherigona spp. Including Sorghum Shoot Fly (*Atherigona soccata* Rond., Muscidae: Diptera). International Crops Research Institute for the Semi-Arid Tropics, Patancheru, Andhra Pradesh, India. 16 pp.

Denmark, H.A., 1988. Sugarcane Aphids in Florida (Homoptera: Aphididae). Entomology Circular No. 302. Division of Plant Industry, Florida Department of Agriculture & Consumer Services.

Denno, R.F., Roderick, G.K., 1990. Population biology of planthoppers. Annual Review of Entomology 35, 489–520.

Dixon, A.F.G., Kindlmann, P., Lepš, J., Holman, J., 1987. Why there are so few species of aphids, especially in the tropics. The American Naturalist 129 (4), 580–592.

Franzmann, B.A., Hardy, A.T., 1996. Testing the host status of Australian indigenous sorghums for the sorghum midge. In: Proceedings Third Australian Sorghum Conference, Tamworth, Occasional Publication No. 93, pp. 365–367.

Franzmann, B.A., Lloyd, R.J., Zalucki, M.P., 2006. Effect of soil burial depth and wetting on mortality of diapausing larvae and patterns of post-diapause adult emergence of sorghum midge, *Stenodiplosis sorghicola* (Coquillett) (Diptera: Cecidomyiidae). Australian Journal of Entomology 45, 192–197.

Getu, E., Overholt, W.A., Kairu, E., 2001. Distribution and species composition of stemborers and their natural enemies in Ethiopia. Insect Science and Its Application 21, 353–359.

Gahukar, R.T., 1991. Recent developments in sorghum entomology research. Agricultural Zoology Reviews 4, 23–65.

Gould, F.W., Shaw, R.B., 1983. Grass Systematics, second ed. Texas A&M University Press.

Hansen, A.A., 1923. Wild corn, a serious weed in Indiana. Indiana Academy of Science Proceedings 33, 295–296.

Harris, K.M., 1961. The sorghum midge, *Contarinia sorghicola* (Coq.) in Nigeria. Bulletin of Entomological Research 52, 129–146.

Harris, K.M., 1979. Descriptions and host ranges of the sorghum midge, *Contarinia sorghicola* (Coquillett) (Diptera. Cecidomyiidae), and of eleven new species of *Contarinia* reared from Gramineae and Cyperaceae in Australia. Bulletin of Entomological Research 69, 161–182.

Jalali, S.K., Singh, S.P., 2001. Studies on the thermal requirements of estimating the number of generations of *Chilo partellus* (Swinhoe) and its natural enemies in the field. Annals of Plant Protection Sciences 9 (2), 213–219.

Jermy, T., 1984. Evolution of insect/host plant relationships. American Naturalist 124, 609–630.

Kawada, K., 1995. Studies on host selection, development and reproduction of *Melanaphis sacchari* (Zehntner). Bulletin of the Research Institute for Bioresources, Okayama University 3, 5–10.

Kfir, R., 1997. Natural control of the cereal stemborers *Busseola fusca* and *Chilo partellus* in South Africa. Insect Science and Its Application 17 (1), 61–67.

Kfir, R., 2002. Increase in cereal stem borer populations through partial elimination of natural enemies. Entomologia Experimentalis et Applicata 104 (2/3), 299–306 50 ref.

Kfir, R., Overholt, W.A., Khan, Z.R., Polaszek, A., 2002. Biology and management of economically important lepidopteran cereal stem borers in Africa. Annual Review of Entomology 47, 701–731.

Khan, S.A., Mulvaney, R.L., Hoeft, R.G., 2000. Direct-diffusion methods for inorganic-nitrogen analysis of soil. Soil Science Society of America Journal 64, 1083–1089.

Liu, Z., Wang, B., Xu, B., Sun, J., 2011. Monoterpene variation mediated attack preference evolution of the bark beetle *Dendroctonus valens*. PLoS One 6, e22005.

Matama-Kauma, T., Schulthess, F., Le Ru, B.P., Mueke, J., Ogwang, J.A., Omwega, C.O., 2008. Abundance and diversity of lepidopteran stemborers and their parasitoids on selected wild grasses in Uganda. Crop Protection 27, 505–513.

Matthee, J.J., 1962. Waak teen plantluise op kafferkoring. Boerdin South Africa 38 (10), 27–29.

Matthee, J.J., Oberholzer, J.J., 1958. Die insekplae op kafferkoring (2). Boerdin South Africa 34 (6), 12–15.

Michels Jr., G.J., 1986. Graminaceous North American host plants of the greenbug with notes on biotypes. Southwestern Entomologist 11, 55–66.

Moran, N., 1986. Benefits of host plant specificity in Uroleucon (Homoptera: Aphididae). Ecology 67, 108–115.

Muller, F.P., Scholl, S.E., 1958. Some notes on the aphid fauna of South Africa. Journal of the Entomological Society of Southern Africa 21, 382–414.

Namba, R., Higa, S.Y., 1971. Host plant studies of the corn planthopper, *Peregrinus maidis* (Ashmead), in Hawaii. Proceedings of the Hawaiian Entomological Society 21, 105–108.

Napompeth, B., 1973. Ecology and Population Dynamics of the Corn Planthopper, *Peregrinus maidis* (Ashmead) (Homoptera: Delphaddae), in Hawaii (Ph.D. dissertation). University of Hawaii, Honolulu, Hawaii. 257 pp.

Nault, L.R., 1983a. Origins in Mesoamerica of maize viruses and mycoplasmas and their leafhopper vectors. In: Plumb, R.T., Thresh, J.M. (Eds.), Plant Virus Epidemiology. Blackwell Scientific Publications, Oxford, England, pp. 259–266. 377 pp.

Nault, L.R., 1983b. Origin of leafhopper vectors of maize pathogens in Mesoamerica. In: Proceedings of the International Maize Virus Diseases Colloquium and Workshop. Ohio Agricultural Research and Development Center, Wooster, Ohio, USA, pp. 75–82.

Nault, L.R., Knoke, J.K., 1981. Maize vectors. In: Gordon, et al. (Ed.), Virus and Viruslike Diseases of Maize in the United States. Southern Cooperative Service Bulletin 247, OARDC, Ohio State Univ., Wooster, Ohio, pp. 77–84.

Nuessly, G.S., Nagata, R.T., 2005. Greenbug, *Schizaphis graminum* (Rondani) (Insecta: Hemiptera: Aphididae). http://entomology.ifas.ufl.edu/creatures.

Nuessly, G.S., Nagata, R.T., Burd, J.D., Hentz, M.G., Carroll, A.S., Halbert, S.E., 2008. Biology and biotype determination of greenbug, *Schizaphis graminum* (Hemiptera: Aphididae), on seashore *Paspalum* Turfgrass (Paspalum vaginatum). Environmental Entomology 37 (2), 586–591.

Nwanze, K.F., Nwilene, F.E., Reddy, Y.V.R., 1998a. Fecundity and diurnal oviposition behaviour of sorghum shoot fly, *Atherigona soccata* Rondani (Diptera: muscidae). Entomon 23 (2), 77–82, 9 ref.

Nwanze, K.F., Nwilene, F.E., Reddy, Y.V.R., 1998b. Evidence of shoot fly *Atherigona soccata* Rondani (Dipt., Muscidae) oviposition response to sorghum seedling volatiles. Journal of Applied Entomology 122 (9/10), 591–594, 14 ref.

Ofomata, V.C., Overholt, W.A., Lux, S.A., Avan, H., Egwuatu, R.I., 2000. Comparative studies on the fecundity, egg survival, larval feeding, and development of *Chilo partellus* and *Chilo orichalcociliellus* (Lepidoptera: Crambidae) on five grasses. Annals of the Entomological Society of America 93 (3), 492–499, 34 ref.

Okech, S.H.O., Neukermans, L.M.N.R., Chinsembu, K.C., 1994. Agroecological distribution of major stalk borers of maize in Zambia. Insect Science and its Application 15, 167–172.

Padmaja, P.G., Madhusudhana, R., Seetharama, N., 2010. Epicuticular wax and morphological traits associated with resistance to shoot fly, *Atherigona soccata* Rondani in sorghum, Sorghum bicolor. Entomon 34 (3).

Paré, P.W., Tumlinson, J.H., 1997. Induced synthesis of plant volatiles. Nature 385, 30–31.

Pitkin, B., Ellis, W., Plant, C., Edmunds, R., 2007. The Leaf and Stem Mines of British Flies and Other Insects. http://www.ukflymines.co.uk/index.php.

Polaszek, A., 1998. African Cereal Stem Borers: Economic Importance, Taxonomy, Natural Enemies and Control. x + 530 pp.; 42 pp. of ref.

Pont, A.C., 1972. The oriental species of *Atherigona* Rondani. In: Jotwani, M.G., Young, W.R. (Eds.), Control of Sorghum Shoot Fly. Oxford & IBH Publishing, New Delhi, India, pp. 27–104.

Porter, D.R., Burd, J.D., Teetes, G., 1997. Greenbug (Homoptera, Aphididae) biotypes, selected by resistant cultivars or preadapted opportunists? Journal of Economic Entomology 90, 1055–1065.

Prestidge, R.A., 1982. The influence of nitrogenous fertilizer on the grassland Auehenorrhyneha (Homoptera). Journal of Applied Ecology 19, 735–749.

Rebe, M., van den Berg, J., McGeoch, M.A., 2004. Colonization of cultivated and indigenous graminaceous host plants by *Busseola fusca* (Fuller) (Lepidoptera: Noctuidae) and *Chilo partellus* (Swinhoe) (Lepidoptera: Crambidae) under field conditions. African Entomology 12 (2), 187–199.

van Rensburg, N.J., van Hamburg, H., 1975. Grain sorghum pests—an integrated control approach. In: Durr, H.J.R., Giliomec, J.H., Neser, S. (Eds.), Proceedings of the First Congress of the Entomological Society of Southern Africa, 30 September–3 October 1974, Stellenbosch, South Africa. Entomological Society of Southern Africa, Pretoria, South Africa, pp. 151–162.

Robbins, J., 1984. Leaf-mining insects in Warwickshire: Part 2. Proceedings of Birmingham Natural History Society 25 (2), 71–88.

Royer, T.A., Pendleton, B.B., Elliott, N.C., Giles, K.L., 2015. Greenbug (Hemiptera: Aphididae) biology, ecology, and management in wheat and sorghum. Journal of Integrated Pest Management 6 (1), 19. http://dx.doi.org/10.1093/jipm/pmv018.

van Rensburg, N.J., 1973. Notes on the occurrence and biology of the sorghum aphid in South Africa. Journal of the Entomological Society of Southern Africa 36, 293–298.

Seshu Reddy, K.V., Davies, J.C., 1977. Species of shoot fly *Atherigona* sp. present in Andhra Pradesh. PANS 23, 379–383.

Setokuchi, O., 1973. Ecology of *Longiunguis sacchari* infesting sorghum. I. Nymphal period and fecundity of apterous viviparous females. Proceedings of the Association for Plant Protection of Kyushu 19, 95–97.

Sharma, H.C., Franzmann, B.A., Henzell, R.G., 2002. Mechanisms and diversity of resistance to sorghum midge, *Stenodiplosis sorghicola*. Euphytica 124, 1–12.

Sharma, H.C., Franzmann, B.A., 2001a. Orientation of sorghum midge, *Stenodiplosis sorghicola*, females (Diptera: Cecidomyiidae) to color and host-odor stimuli. Journal of Agricultural and Urban Entomology 18 (4), 237–248.

Sharma, H.C., Franzmann, B.A., 2001b. Host- plant preference and oviposition response of the sorghum midge, *Stenodiplosis sorghicola* (Coquillett) towards wild relatives of sorghum. Journal of Applied Entomology 125, 109–114.

Sharma, H.C., Leuschner, K., Vidyasagar, P., 1990. Factors influencing oviposition behaviour of the sorghum midge, *Contarinia sorghicola* Coq. Annals of Applied Biology 116, 431–439.

Sharma, H.C., Vidyasagar, P., Leuschner, K., 1988. Nochoice cage technique to screen for resistance to sorghum midge (Cecidomyiidae: Diptera). Journal of Economic Entomology 81, 415–422.

Shufran, K.A., Peters, D.C., Webster, J.A., 1997. Generation of clonal diversity by sexual reproduction in the greenbug, *Schizaphis graminum*. Insect Molecular Biology 6, 203–209.

Shufran, K.A., Burd, J.D., Anstead, J.A., Lushai, G., 2000. Mitochondrial DNA sequence divergence among greenbug (Homoptera: Aphididae) biotypes: evidence for host-adapted races. Insect Molecular Biology 9, 179–184.

Shyam Prasad, G., Elangovan, M., Bhagwat, V.R., Kalaisekar, A., Subbarayudu, B., 2010. Assortment of indigenous sorghum [*Sorghum bicolor* (L.) Moench] germplasm based on reaction to stem borer, *Chilo partellus* (Swinhoe) under artificial infestation. Indian Journal of Plant Protection 38 (2), 126–130.

Shyam Prasad, G., Bhagwat, V.R., Kalaisekar, A., Subbarayudu, B., Umakanth, A.V., Kannababu, N., 2011. Assessment of resistance to stem borer, *Chilo partellus* (Swinhoe) in sweet sorghum, [*Sorghum bicolor* (L.) Moench]. Indian Journal of Entomology 73 (2), 116–120.

Shyam Prasad, G., Bhagwat, V.R., Srinivasa Babu, K., Kalaisekar, A., Subbarayudu, B., 2015. Identification of forage sorghum lines having multiple-resistance to sorghum shoot fly and spotted stem borer. Range Management & Agroforestry 36 (2), 164–169.

Singh, B.U., Seetharama, N., 2008. Host plant interactions of the corn planthopper, *Peregrinus maidis* Ashm. (Homoptera: Delphacidae) in maize and sorghum Agroecosystems. Arthropod-Plant Interactions 2, 163–196.

Skidmore, P., 1985. The Biology of the Muscidae of the World. Kluwer Academic Publisher, 3300 AH Dordrecht, The Netherlands. 550 pp.

Subbarayudu, B., Shyam Prasad, G., Kalaisekar, A., Bhagwat, V.R., Elangovan, M., 2011. Evaluation of sorghum cultivars for multiple resistance to shoot pests. Indian Journal of Plant Protection 39 (2), 116–120.

Sujatha, N.K., Pape, T., Pont, A.C., Wiegmann, B.M., Meier, R., 2008. The Muscoidea (Diptera: Calyptratae) are paraphyletic: evidence from four mitochondrial and four nuclear genes. Molecular Phylogenetics and Evolution 49 (2), 639–652.

Sujatha, N.K., Pont, A.C., Meier, R., Pape, T., 2014. Complete tribal sampling reveals basal split in Muscidae (Diptera), confirms saprophagy as ancestral feeding mode, and reveals an evolutionary correlation between instar numbers and carnivory. Molecular Phylogenetics and Evolution 78, 349–364.

Swami, H., Bajpai, N.K., 2006. Preference of graminaceous host plants to stem borer, *Chilo partellus* (Swinhoe) at Udaipur, India. Journal of Plant Protection and Environment 3 (2), 67–71.

Teetes, G.L., 1985. Sorghum midge biology, population dynamics and integrated pest management. In: Proc Int Workshop Sorghum Insect Pests. Texas A&M University and ICRISAT, pp. 233–245.

Tsai, J.H., 1996. Development and oviposition of *Peregrinus maidis* (Homoptera: Delphacidae) on various host plants. Florida Entomologist 70, 19–26.

Tsai, J.H., 1975. Occurrence of a corn disease in Florida transmitted by *Peregrinus maidis*. Plant Disease Reporter 59, 830–833.

Tsai, J.H., Falk, B.W., 1988. Tropical maize pathogens and their associated insect vectors. In: Harris, K.F. (Ed.), Advances in Disease Vector Research, vol. 5. Springer-Verlag, New York, NY, pp. 177–201.

Visser, J.H., 1986. Host odor perception in phytophagous insects. Annual Review of Entomology 31, 121–144.

Walter, E.V., 1941. The biology and control of sorghum midge. Technical Bulletin 778, USDA, Washington D.C., p. 27.

Webster, F.M., Phillips, W.J., 1912. The spring grain aphis or "greenbug". U.S. Burkness Entomological Bulletin 110, 1–153. https://archive.org/details/springgrainaphis110webs.

Weng, Y., Perumal, A., Burd, J.D., Rudd, J.C., 2010. Biotypic diversity in greenbug (Hemiptera: Aphididae): microsatellite-based regional divergence and host-adapted differentiation. Journal of Economic Entomology 103 (4), 1454–1463.

Whitcomb, R.F., Kramer, J., Cuan, M.E., Hicks, A.L., 1987. Ecology and evolution of leafhopper-grass host relationships in North American grasslands. In: Harris, K.F. (Ed.), Current Topics in Vector Research. Springer-Verlag, New York, pp. 121–178.

Wilbrink, G., 1922. An investigation on spread of the mosaic disease of sugarcane by aphids. Medid Procfst Java Suikerind 10, 413–456.

Yadava, R.L., 1966. Oviparity in sugarcane aphid, *Longiunguis sacchari* Zehnt. (Aphidae: Homoptera). Current Science 1, 18.

Zhang, Q.H., Schlyter, F., 2004. Olfactory recognition and behavioural avoidance of angiosperm nonhost volatiles by conifer-inhabiting bark beetles. Agriculture and Forest Entomology 6, 1–19.

FURTHER READING

Singh, B.U., Padmaja, P.G., Seetharama, N., 2004. Biology and management of the sugarcane aphid, *Melanaphis sacchari* (Zehntner) (Homoptera: Aphididae), in sorghum: a review. Crop Protection 23, 739–755.

PEST MANAGEMENT STRATEGIES AND TECHNOLOGIES

HOST-PLANT RESISTANCE

A major component of integrated pest management strategies in cereals is the use of host-plant resistance. Host-plant resistance is an effective, economical, and environmentally friendly approach to managing insect pests. Painter (1951) defined plant resistance to insects as "the heritable characteristics possessed by a plant which influence the ultimate degree of damage done by an insect."

MECHANISMS

Three different mechanisms of insect resistance, namely nonpreference, antibiosis, and tolerance, have been observed to be involved in insect–host-plant interactions (Painter, 1951). The term "nonpreference" was subsequently replaced with "antixenosis" (Kogan and Ortman, 1978). Antixenosis detrimentally affects insects as they attempt to use plants for food, ovipositioning, or shelter. For example, globrous or hairy leaf surface could be non-preferred plant trait for some insects such as white flies, aphids, etc. The resistant plant is then rejected by the pest as an unsuitable host. Therefore, obviously antixenosis is the most preferred mechanism of resistance due to its ability to prevent the insect attack. The antibiosis mechanism of resistance occurs after host-plant infestation whereby the biology of the insect is affected as it feeds on the plant. Tolerance is the ability of the plant to recover from the insect damage. Resistant plants may contain one or a combination of the three mechanisms that collectively contribute to the level of insect resistance. Each mechanism of resistance acts at some stage of the insect–plant relationship and is contributed to by physical or chemical plant characters that may be referred to as components of resistance. Nonpreference or antixenosis mechanisms of resistance occur during the colonization phase. Antibiosis resistance occurs at the utilization phase of the insect–plant interaction and tolerance mechanism operates after the insect attack. In a nutshell, antixenosis, antibiosis and tolerance mechanisms operate *before, during* and *after* the insect attack respectively. The plant traits that confer host plant resistant to insect pests in sorghum are presented in Table 5.1.

OVIPOSITIONAL ANTIXENOSIS/NONPREFERENCE

Nonpreference by insects is the property of the plant to render it unattractive for ovipositioning, feeding, or shelter. The physicochemical characteristics of the host plant affect insect behavior and make the host unsuitable for the survival. The absence of physicochemical stimuli that are involved in the selection of host plant or the presence of repellents, deterrents, and antifeedants contribute to the antixenosis mechanism of resistance. Ovipositioning nonpreference is considered to be a primary mechanism of resistance to major insect pests of sorghum, viz., shoot fly, head bug, and midge

Table 5.1 Specific Traits Associated With Mechanisms of Host Plant Resistance Against Various Pests in Sorghum

Resistance Mechanism	Pest	Associated Traits/Characters in Sorghum
Antixenosis	Shoot fly, *Atherigona soccata*	Low transpiration rate (Mate et al., 1988); leaf surface wetness (Nwanze et al., 1990); epicuticular wax (Nwanze et al., 1992); leaf glossiness (Blum, 1972); trichomes on the abaxial leaf surface (Blum, 1968; Maiti et al., 1980); seedling vigor (Taneja and Leuschner, 1985)
	Stem borer, *Chilo partellus*	Oviposition nonpreference, reduced leaf feeding, low deadheart formation, reduced tunneling (Dabrowski and Kidiavai, 1983; Woodhead and Taneja, 1987; Sharma and Nwanze, 1997)
	Shoot bug, *Peregrinus maidis*	Compact and tight leaf whorl (Agarwal et al., 1978)
	Sorghum midge, *Stenodiplosis sorghicola*	Glume size and extent of glume closure (Bergquist et al., 1974; Jadhav and Jadhav, 1978; Rossetto et al., 1975); glume, palea, lemma, anther, and style length (Sharma, 1985; Sharma et al., 1990a,b, 2002)
Antibiosis	Shoot fly, *A. soccata*	High enzyme activities, especially of peroxidase and polyphenol oxidase (Patil et al., 2006); deposition of irregularly shaped silica bodies, distinct lignification and thickening of cell walls (Ponnaiya, 1951; Blum, 1968)
	Stem borer, *C. partellus*	Low sugar content; greater amounts of amino acids, tannins, total phenols, neutral detergent fiber, acid detergent fiber, lignins, and silica content (Swarup and Chaugale, 1962; Narwal, 1973; Khurana and Verma, 1982, 1983; Torto et al., 1990)
	Shoot bug, *P. maidis*	Increased mortality, prolonged nymphal development, and reduced fecundity (Chandra Shekar, 1991)
	Sorghum midge, *S. sorghicola*	Decreased rates of postembryonic growth, survival, and adult fecundity; larvae smaller in size and in weight (Sharma et al., 1993; Wuensche, 1980); failure of larvae to pupate (Waquil et al., 1986); delayed adult emergence from pupa, decreased fecundity, and lower rates of progeny production (Sharma et al., 1993); shortened postembryonic life cycle (Natarajan and Chelliah, 1985); higher larval mortality (Rossetto, 1977; Sharma et al., 1993)
	Head bug, *Calocoris angustatus*	Extended postembryonic development, lower weights in nymphs and adults, smaller numbers of nymphal survivors, reduced food utilization efficiency at nymphal stage and reduced food consumption index and growth rate (Sharma and Lopez, 1990)
Tolerance	Shoot fly, *A. soccata*	High plant recovery (Sharma et al., 1977); high rate of tiller survival, faster growth of tiller, high rate of tiller growth (Blum, 1972); fewer deadhearts in tillers
	Stem borer, *C. partellus*	Tolerance to leaf damage and stem tunneling (Sharma and Nwanze, 1997)
	Sorghum midge, *S. sorghicola*	Positive levels of yield compensations (Sharma et al., 2002)

(Blum, 1967; Sharma and Lopez, 1990; Sharma and Vidyasagar, 1994). There are several plant traits postulated to provide antixenotic effects against the potential pest species.

Shoot Flies

Nonpreference is an important mechanism of resistance to the shoot fly, *Atherigona soccata* in sorghum. Nonpreference for ovipositioning in sorghum is relative, because none of the known resistant cultivars were completely nonpreferred for egg-laying. There are several parameters or traits in sorghum associated with oviposition non preference for *A. soccata* identified. Sorghum genotypes with high transpiration rates are preferred for ovipositioning (Mate et al., 1988). This may be due to the creation of favourable microclimatic humidity under high transpiration. Leaf surface wetness along with epicuticular wax is associated with susceptibility, and leaf glossiness with resistance, to shoot fly (Blum, 1972; Maiti et al., 1984; Agrawal and Abraham, 1985; Nwanze et al., 1990, 1992; Kamatar and Salimath, 2003). The genotypes with glossy leaves during seedling stage are found to be less preferred for oviposition and thus show less number of deadhearts (Sharma et al., 1997). The seedlings that are less preferred by the shoot fly for oviposition are with higher density of trichomes (Blum, 1968; Maiti et al., 1980). Trichome density and morphology (Fig. 5.1) (unicellular trichomes) and plant resistance to shoot fly have a positive association (Gibson and Maiti, 1983; Omori et al., 1983; Padmaja et al., 2010). Trichomeless cultivars accumulate more dew and stay wet longer facilitating the movement of freshly hatched larvae to the base of central shoot. On the other hand, trichomed cultivars would tend to dry faster, making the downward journey of the larvae more difficult (Raina et al., 1981). Although trichome density is significantly and negatively correlated with deadhearts, it does not have direct role in reducing deadhearts, but contributes to shoot fly resistance mainly through other traits (Karanjkar et al., 1992). There was a higher level of resistance to shoot fly when leaf glossiness and trichomes occurred together in a genotype. Resistant genotypes have a smooth, amorphous wax layer with few wax crystals, whereas susceptible genotypes have significantly more wax in the epicuticle (Fig. 5.2).

(A)   **(B)**

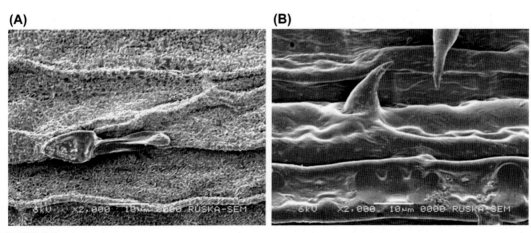

FIGURE 5.1

Variation in trichome morphology in sorghum genotypes. (A) Bicellular trichome in shoot fly susceptible, 296B. (B) Unicellular, pointed trichome in shoot fly resistant, IS 18551.

Courtesy of P.G. Padmaja.

(A)   **(B)**

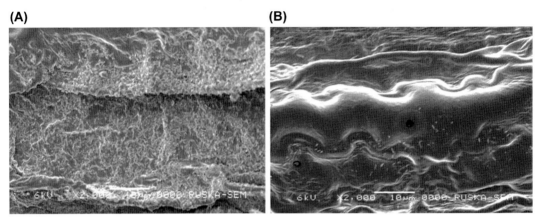

FIGURE 5.2

Variation in epicuticular wax structure in sorghum genotypes. (A) Dense wax crystals in shoot fly susceptible, 296B. (B) Smooth amorphous wax in shoot fly resistant, IS 18551.

Courtesy of P.G. Padmaja.

The other traits such as frequency of lines with high vigor score is also found to show resistance in terms of lesser number of eggs laid (Taneja and Leuschner, 1985; Sharma et al., 1997) however, seedling vigour has not been a consistent trait across a wide spectrum of genotypes (Dhillon, 2004). The role of trichomes in resistance to shoot fly is questionable. The proponents of trichome theory have based their hypothesis on oviposition non preference by female flies and they seem to explain that the trichomes could impede with larval movement on leaf. On the contrary, *Atherigona* species are able to survive and produce deadhearts on several grasses with densely globrus and hairy leaf surfaces. Although trichome density is significantly and negatively correlated with deadhearts, it does not have a direct role in reducing deadhearts, but contributes to shoot fly resistance mainly through other traits (Karanjkar et al., 1992). Plants with eggs, deadhearts, leaf glossiness, trichomes on the abaxial surface of the leaf, and leaf sheath pigmentation are the most reliable parameters, and these can be used as marker traits to screen and select for resistance to sorghum shoot fly (Blum, 1968; Maiti et al., 1980; Gibson and Maiti, 1983; Omori et al., 1983; Taneja and Leuschner, 1985; Sharma et al., 1997; Dhillon, 2004).

Stem Borers

There have been several components of resistance to *C. partellus* in sorghum identified, including oviposition nonpreference (Singh and Rana, 1984; Alghali, 1985; Saxena, 1990; van den Berg and van der Westhuizen, 1997), reduced leaf-feeding, less number of deadhearts, and reduced stem tunneling (Dabrowski and Kidiavai, 1983; Woodhead and Taneja, 1987; Sharma and Nwanze, 1997).

Shoot Bug and Other Bugs

Antixenosis works against planthopper, *Perigrinus maids* in sorghum with compact and tightly wrapped whorl leaves around the stem (Agarwal et al., 1978). This mechanism impede with colonization, oviposition, and feeding (Chandra Shekar, 1991; Chandra Shekar et al., 1992, 1993a,b).

Levels of nitrogen, sugar and total chlorophyll contents influence the colonization of *P. maidis*. In general, the ability of the planthoppers to synchronize the nitrogen levels of host plants is attributed for the host specialization (Denno and Roderick, 1990). Therefore, the nutrient levels in the plants are very important in regulation of populations of bugs.

High phosphorus, potash, and polyphenol content are less preferred by *P. maidis* (Mote and Shahane, 1994).

The antixenosis mechanism of resistance may be closely linked with the structural morphology of spikelets small glume size and the extent of glume closure and the length of glume, palea, lemma, anther, and style (Bergquist et al., 1974; Jadhav and Jadhav, 1978; Rossetto et al., 1975; Rossetto et al., 1984; Sharma, 1985; Franzmann, 1988; Waquil et al., 1986; Henzell et al., 1994; Sharma and Vidyasagar, 1994; Sharma et al., 1990a,b, 2002). Glumes of spikelets of resistant varieties were more tightly closed than those in susceptible varieties (Diarisso, 1997). *Sorghum halepense* is also a preferred sorghum species apart from *S. bicolor* whereas the other sorghum species such as *S. angustum, S. amplum, S. bulbosum, S. brachypodum, S. laxiflorum, S. macrospermum, S. nitidium,* and *S. stipoideum are not suitable for oviposition* (Sharma and Franzmann, 2001). Therefore, the wild species of sorghum can be used as sources of midge resistant genes (Sharma and Franzmann, 2001).

ANTIBIOSIS

Antibiosis is the result of action of plant-biochemicals in the biological processes of herbivorous insects. Generally, it is expressed in terms of larval mortality, decreased larval and pupal weights, prolonged larval and pupal development, reduced fecundity, and prolonged generation time and overall effect on insect survival and development.

Shoot Fly

In sorghum shoot fly, *A. soccata*, reduced growth and development, prolonged durations of larval and pupal periods, and poor or failure in adult emergence on resistant genotypes are some of the pronounced effects of antibiosis (Singh and Jotwani, 1980b; Raina et al., 1981; Sharma et al., 1997). Antibiosis of shoot fly offers exciting possibilities of exerting biotic pressure against insect feeding and development, resulting in low-larval survival on resistant varieties (Dahms, 1969; Soto, 1974). However, the antibiosis as a host-plant resistant mechanism against shoot fly needs a thorough investigation in light of the larval feeding habit being saprophagy. Nevertheless, there are several instances of antibiosis against shoot fly in sorghum exist in the form of high mortality at first instar (Zein el Abdin, 1981), low-larval survival (Dahms, 1969; Soto, 1974; Ogwaro and Kokwaro, 1981), low total growth indices (Dhawan et al., 1993; Dhillon et al., 2005), Resistance to shoot fly is a cumulative effect of non-preference and antibiosis (Raina et al., 1981). Survival and development were adversely affected when shoot flies were reared on resistant cultivars (Jotwani and Srivastava, 1970; Narayana, 1975; Singh and Narayana, 1978; Singh and Jotwani, 1980a; Raina et al., 1981; Unnithan and Reddy, 1985; Sharma et al., 1997; Dhillon et al., 2005). Reduced fecundity, prolongation in larval and pupal periods, and lower larval survival (Singh and Narayana, 1978; Singh and Jotwani, 1980a; Taneja and Leuschner, 1985; Jadhav and Mote, 1986; Dhillon et al., 2005) and lower percentage pupation on resistant cultivars (Ogwaro and Kokwaro, 1981; Dhawan et al., 1993; Dhillon et al., 2005) were attributed to antibiosis mechanism of resistance in sorghum. Resistance to shoot fly is a cumulative effect of non-preference and antibiosis (Raina et al., 1981).

Biochemical deficiencies or the presence of chemical factors in resistant cultivars might adversely affect the development and survival of shoot fly larvae (Raina, 1985). High enzyme activities especially of peroxidase and polyphenol oxidase occurs in resistant lines as well as resistant×resistant and resistant×susceptible crosses (Patil et al., 2006). The higher enzyme activity might be inducing and activating the antibiosis mechanism, leading to reduction in damage caused by the shoot fly. Antibiosis was also attributed to early deposition of irregular-shaped silica bodies such as dumbbell-shaped, intercostal and silicified prickle hairs, in the abaxial epidermis of the leaf sheaths and distinct lignification and thickening of walls of the cells enclosing the vascular bundle sheaths within the central whorl of young leaves (Ponnaiya, 1951; Blum, 1968). The greater density of silica bodies in the resistant cultivars offers protection against shoot fly damage and the density of such silicate structures increases from first to third leaf-sheath (Blum, 1968).

Stem Borer
Both tolerance and antibiosis are operating in resistant cultivars (Jotwani, 1976). The secondary plant substances in the leaves and poor nutritional quality of the resistant cultivars of the host-plant hinder the normal development of *C. Partellus*. Low sugar content, greater amounts of amino acids, tannins, total phenols, neutral detergent fiber, acid detergent fiber, lignins and silica content (Swarup and Chaugale, 1962; Narwal, 1973; Khurana and Verma, 1982, 1983; Torto et al., 1990) are associated with resistance to *C. partellus* in sorghum. Some of the pronounced effects of antibiosis on *C. partellus* are prolonged larval, pupal and total development periods, and reduced pupal weight, low pupation and adult emergence (Lal and Sukhani, 1982; Singh and Verma, 1988; Saxena, 1992; Verma et al., 1992).

Aphids
Antibiosis is the principal mechanism that operates against aphids, *Melanaphis sacchari* in sorghum and the reduced population/cm^2, aphid damage rating, chlorophyll content and grain yield (g)/plant are the most reliable parameters for characterization of resistant sorghum cultivars to *M. sacchari* (Bhagwat et al., 2011).

Shoot Bug
Antibiosis is expressed as increased mortality, prolonged nymphal development, and reduced fecundity in *P. maidis* (Chandra Shekar, 1991) feeding on resistant genotypes of sorghum.

Midge
Antibiosis to midge in sorghum leads to decreased rates of postembryonic growth, survival, and adult fecundity. Larvae developing on resistant cultivars become smaller in size and in weight (Sharma et al., 1993; Wuensche, 1980). Successful development of larvae in to pupae is hampered on several resistant cultivars of sorghum (Waquil et al., 1986). The larval death due to antibiosis occurs rarely in sorghum before the development of embryo of the floret in to seed (Sharma, 1985). This the most desirable antibiosis effect that needs to be further investigated to breed better resistant cultivars. There are several other effects of antibiosis observed such as delayed adult emergence from pupa, decreased fecundity and lower rates of progeny production (Sharma et al., 1993), shortened postembryonic life cycle (Natarajan and Chelliah, 1985), higher larval mortality in midge resistant lines (Rossetto, 1977; Sharma et al., 1993). Midge resistance lines of sorghum found to have higher grain (caryopsis) growth rates, and increased tannin contents (Sharma, 1985, 1993; Sharma et al., 1993). However, the relation between

increased tannin content and midge resistance does not hold good with reference to sorghum cultivar DJ6514, which shows very high levels of antibiosis against the midge in terms of high rates of mortality of the immature stages of midge development (Sharma et al., 1993).

Head Bug

Similar to other insect pests, *Calocoris angustatus* experienced disturbances in developmental processes. Extended post-embryonic development, lower weights in fifth instar nymphs and adults, lesser numbers of nymphal survival, reduced food utilization efficiency at fourth instar nymphal stage and reduced food consumption index and growth rate (Sharma and Lopez, 1990) are some of the important indicators of antibiosis effect in *C. angustatus* when feeding on resistant cultivars of sorghum.

TOLERANCE/RECOVERY RESISTANCE

Tolerance is also called recovery resistance is the ability of the plant to withstand or recover from damage caused by insect. In case of shoot fly, some genotypes of sorghum are able to produce tillers when the main shoot is killed, which are more resistant to shoot fly attack (Blum, 1969; Dogget et al., 1970; Dogget, 1972). High-plant recovery (Sharma et al., 1977), high rate of tiller survival, faster growth of tiller, high rate of tiller growth (Blum, 1972) are some of the characteristics of resistant varieties. Recovery resistance may not have definite relation with height of the plant, as some of the tolerant germplasm lines are dwarf, medium tall or very tall (Shivankar et al., 1989; Dhillon, 2004). Higher yield under shoot fly infestation is also considered as recovery resistance (Rana et al., 1985). Shoot fly tolerance can be seen in genotypes with significantly less deadhearts in tillers than in tillers of the susceptible ones. However, the indicators of tolerance or recovery resistance are highly dependant on the level of primary resistance and shoot fly abundance (Dogget et al., 1970; Dhillon, 2004). Recovery resistance does not appear to be useful mechanisms of resistance particularly when shoot fly population increases progressively as the rainy season continues (Singh and Rana, 1986). During 1960s and 1970s, the African sorghum cultivars such as *Serena* and *Namatrare* were able to recover when more than 90% of the main shoots of the plants were killed by shoot fly attack (Dogget and Majisu, 1965; Dogget et al., 1970). In Africa, it was reported that farmers actually preferred an initial infestation of sorghum by shoot fly that led to profuse tillering and subsequently a good harvest (Dogget, 1972). The damaged plants produce axial tillers, which serve as a mechanism of recovery resistance. But, the productivity of side tillers may not be matching with that of main shoot. Because, the axial tillers often mature later than the main plants and often suffer greater damage by sorghum midge, Stenodiplosis sorghicola, head bugs, Calocoris angustatus and birds or may not be able to produce grain under drought stress (Dhillon, 2004). However, tolerance can be greatly influenced by the growth conditions of the plant and thus may not always be predictable at various locations, particularly those with irregular patterns of rainfall (Raina, 1985).

Tolerance to sorghum midge, *S. sorghicola* has not been well established in sorghum. There are some levels of compensation occur in certain genotypes in terms of seed set, but not to scale of significantly separable as resistance and susceptible (Franzmann and Butler, 1993). Positive levels of yield compensations occurred in several genotypes of sorghum that experienced midge attack and such genotypes can be grouped as tolerant/resistant lines (Sharma et al., 2002). However, sorghum has the ability to compensate yield within panicles when spikelets or developing kernels are physically removed (Fisher and Wilson, 1975; Hamilton et al., 1982; Henzell and Gillieron, 1973). Therefore, existence of specific tolerance to sorghum midge among genotypes requires further exploration and understanding.

SOURCES OF INSECT RESISTANCE

Millets have a wide diversity of genetic pool in terms of several traits such as yield, resistance to pests and diseases, etc. Exploring the millets-diversity for identifying insect pest resistant sources has been a major activity in cultivar development programmes across Asia and Africa. In India, a huge network of scientist has been working on systematically screening world germplasm collections through All India Coordinated Research Projects (AICRP) on Sorghum and Small Millets. International Crops Research Institute for Semi-Arid Tropcs (ICRISAT) located in India has also been undertaking extensive screening of the sorghum germplasm collections for resistance to key sorghum pests such as sorghum shoot fly, spotted stem borer, sorghum midge and head bugs and resistant lines for shoot fly and other pests have been identified (Sharma et al., 2008). There are over 36,700 sorghum germplasm accessions in ICRISAT genebank which serves as a global repository of the sorghum germplasm (Sharma et al., 2008). Many of the resistant sources identified in the germplasm are poor combiners for yielders. Therefore breeding for cultivars with high yield and insect resistance has been a tough task. Identified resistance sources mostly come from *maldandi* (semi-compact head type) or *dagadi* (compact head type) races grown in post-rainy season in India (Rana et al., 1985). *Maldandi* is the popular landrace that has been grown by farmers mostly in the regions of Maharashtra, India during post-rainy season for grain purpose (Rana et al., 1985; Rakshit et al., 2012; Sanjana and Patil, 2015).

In India, much progress has been made in identifying resistant sources and in understanding the mechanisms of resistance to insect pests of sorghum, foxtail millet, kodo millet and finger millet (Table 5.2).

The existence of genetic differences in sorghum for resistance to shoot fly was first established by Ponnaiya (1951) in India. Identification of resistance sources for shoot fly resistance in sorghum has been a daunting task since 1960. There are variations in damage percentages among different genotypes, but no genotype has ever shown immunity to the shoot fly attack so far (Singh et al., 1968, 1978; Sharma, 1997). Identified resistance sources mostly come from *maldandi* (semi-compact head type) or dagadi (compact head type) races grown in post-rainy season in India (Rana et al., 1985). Maldandi is the popular landrace that has been grown by farmers mostly in the regions of Maharashtra, India during post-rainy season for grain purpose (Rana et al., 1985; Rakshit et al., 2012; Sanjana and Patil, 2015). Wild species of sorghum such as *Sorghum purpureosericeum, S. versicolor, Parasorghum (S. australience, S. purpureosericeum, S. brevicallosum, S. timorense, S. versicolor, S. matarankense and S. nitidum) and Stiposorghum (S. angustum, S. ecarinatum, S. extans, S.intrans, S. interjectum and S. stipoideum), Heterosorghum (S. laxiflorum) and Chaetosorghum (S. macrospermum)* offer very high levels of resistance to shoot fly (Mote, 1984; Venkateswaran, 2003). Combinations of resistant mechanisms operate in wild relatives of sorghum against shoot fly. For example, IS 18226 (*race arundinaceum*) and IS 14212 (*S. halepense*) reduce survival and fecundity in *A. soccata*. Wild relatives of sorghum exhibit very high levels of antibiosis to shoot fly. Therefore, wild relatives of sorghum are the suitable candidates as a source of alternate genes to increase the resistance in cultivated sorghum A. soccata (Kamala et al., 2009).

There is huge collection of germplasm lines identified for shoot fly resistance. As far as shoot fly is concerned, the word resistance is purely comparative in the sense that the resistant genotypes are relatively better than the susceptible ones. Several lines have been identified and developed to offer certain level of resistance to shoot fly. Some of the important genotypes are IS 2123, SPSFR 94006, SPSFR 94007, SPSFR 94011, SPSFR 94019, SPSFR 94034, SPSFR 96069, SPSFR 86065, PS 23585, ICSV

Table 5.2 Resistant Sources Identified for Various Insect Pests in Millets

Pest	Crop and Names of Resistant Genotypes
Shoot fly, *Atherigona* spp.	Sorghum
	IS 2123, ICSV 705, ICSV 708, SPSFR 94019, SPSFR 94006, SPSFR 94007, SPSFR 94011, SPSFR 94034, ICSV 93127, SPSFR 96069, SPSFR 86065, PS 23585, ICSR 89058; PBMR3, PBMR7, PBMR8, BMR23375, BMR23150, DSRBMR1 (BMR lines) and (RS4007 x IS3691)-1-1-1-1, (279B x11B2)-ab5 pl 1-1-1-1, (11B2 x RS2309-1B2)-1-1-1-1, (ICSB51 X 11B2)-2-1-1-1-1, Chittapur Local, EP33, PS54, PS164, PS219, RSSV9, NRCSFR09-3, GMR309, BS8586, and ICSV700, ICSV93046, IC2123, IS2146 (IIMR, 2016); Pirira-1, Pirira-2, Sima, SV-1, Larsvyt46-85, SDSL 87046, SDSL89473, Mmabaitse, SV-2, SDSL 89473, ZSV-15, Larsvyt46-85, Macia, SDSL 98014, Kuyuma (van den Berg et al., 2005)
	Pearl millet
	MH 1078, MH 1099, MH 1139, MH 1079, Saburi, Raj 171, Pusa 383, ICTP 8203, MH 1038, MH 1049, MH 1109, and Pusa 266 (ICAR, 2002–07)
	Foxtail millet
	SIA 1538, SIA 1533, SIA 1507, SIA 1581, SIA1566, SIA 1549
	Kodo millet
	RPS 40-1, RPS 40-2, RPS 62-3, RPS 72-2, RPS 120-1, IPS 6, IPS 32, IPS 110, IPS 131, IPS 142, IPS 178, Keharpur (Murthy and Harinarayana, 1989). IPS 147-1, JK 41, RPS 140-1, RPS 136-1 (Singh et al., 1990)
Stem borer, *Chilo partellus*	Sorghum
	IS 1044, IS 1054, IS 2123, IS 2263, IS 2269, IS 5469, IS 5566, IS 12308, IS 13100, IS 18333, and IS 18573) (Sharma et al., 2003); IS 18584, IS 18577 and IS (Patil et al., 1996); P-217, P-297, P-500, P-291, P-84, P-296, P-467, P-471 and P-495, E-303, P-217, P-297, and P-500 (Kishore, 1987); KC-1, PGN-1, PGN-20 and PGN-64, PGN-1, PGN-64, PGN-20, AKENT-20, and KC-1 (Kishore, 2001); SPV 1518, SPV 1489, SPV 462, SPH 1148, SPH 1270, SPH 1280, CSH 17, SPV 1572, SPV 1563, SPV 1565, CSH 16, SPH 1335, and CSV 15 (Kishore et al., 2002)
	Pearl millet
	MH 1140, Saburi, MP 430, Raj 171, ICMB 221, MH1038, MH 1049, MH 1109, MH 1152, MP 414, RCB 2, MH–1142, MH–1188, MH–1192, MH–1194, MH–1195, H–1199, MH–1201, MH–1206, MH–1213, MH–1218, and MH–1139 (ICAR, 2002–07)
Sugar cane aphid, *Melanaphis sacchari*	Sorghum
	HB 37, PE 954177, IS 8100C, R128, R131, and R133 (Sharma, 1993); SPS43, SLR37, TAM428, SLB81, KR191, Long SPS43, and SLR37 (Bhagwat et al., 2011)
Corn aphid, *Rhopalosiphum maidis*	Sorghum
	Piper Sudan 428-1, CS 3541, and TAM 428 (Sharma, 1993)
Green bug, *Schizaphis graminum*	Sorghum
	PI 302178, PI 302236, IS 809, EA 71, EA 226, EA 252, Kafir 60 x H 39, and H 39 (Harvey and Hackerott, 1974; Sharma, 1993).
Sorghum midge, *Stenodiplosis sorghicolaa*	Sorghum
	IS 3461, IS 9807, IS 10712, IS 18563, IS 19476, IS 21873, IS 21881, IS 22806, PM 15936-2, and ICSV 197 (Sharma et al., 2002)

Continued

Table 5.2 Resistant Sources Identified for Various Insect Pests in Millets—cont'd

Pest	Crop and Names of Resistant Genotypes
Shoot bug, *Peregrinus maidis*	Sorghum Kafir Suma and Dwarf Hegari (Khan and Rao, 1956); I 753, H 109, GIB, 3677B, and IS 1055 (Agarwal et al., 1978)
Head bug, *Calocoris angustatus*	Sorghum IS 17645, IS 21443, IS 17618, CIS 17610, IS 2761, IS 9692, IS 9639, IS 19940, IS 19950, IS 19957, IS 25760, IS 21444 (Sharma, 1993)
Head bug, *Eurystylus immaculatus*	M 388, S 29, IS 14332 (Sharma, 1993)
Sugar cane leafhopper, *Pyrilla perpusilla*	Pearl millet MH–1121, Saburi, CZP–9082 (ICAR, 2002–07)
Gray weevil, *Myllocerus* spp.	Pearl millet MH–1187, MH–1153, MH–1205, MH–1231, MH–1236, and RHB–121 (ICAR, 2002–07)
Chinch bug, *Blissus* spp.	Pearl millet 07F-1226, 07F-1229, 07F-1231, 07F-1235, 07F-1238, 07F-1239, and 07F-1240 (Ni et al., 2009)

705, ICSV 708, ICSV 93127, ICSR 89058. Some of the sweet stock lines resistant to shoot fly are IS 2205, IS 18573, IS 5448, IS 5470, ICSV 700 and ICSV 93046 are used in crop improvement in Asia, Africa, USA, and Australia.

Plants with eggs, deadhearts, leaf glossiness, trichomes on the abaxial surface of the leaf, and leaf sheath pigmentation are the most reliable parameters, and these can be used as marker traits to screen and select for resistance to sorghum shoot fly.

Stem Borer

The resistance to stem borer in cultivated sorghum genotypes are more pronounced and significant. There are several genotypes identified as donors of resistant genes for stem borer such as IS 18577, IS 18584, and IS 2205, P-84, P-217, P-297, P-291,P-296, P-467, P-471, P-495 and P-500 (Kishore, 1987). Generally, the resistance to *Chilo partellus* is assessed on the basis of dead hearts, leaf injury, stem tunnelling, peduncle tunnelling and exit holes (Patil et al., 1996).

The resistant genotypes offer antibiosis in the form of differences in adult longevity, variations in the developmental periods of *C. partellus* on different cultivars and are attributed to the nutritional value of the food source (Singh and Marwaha, 1996). The resistant genotype IS 18551 reduces the growth index in *C. partellus* showing an antibiosis reaction as compared to susceptible cultivars CSH-1 and CSH-9 (Singh and Marwaha, 1996). There are several germplasm lines showing resistance to stem borer along with good agronomical traits and yield viz., SPV 1518, SPV 1489, SPV 462, SPH 1148, SPH 1270, SPH 1280, CSH 17, SPV 1572, SPV 1563, SPV 1565, CSH 16, SPH 1335 and CSV 15, PGN-1, PGN-64, PGN-20, AKENT-20 and KC-1 (Kishore, 2001; Kishore et al., 2002).

There are some genotypes which show multiple resistances to more than one insect pest. For example, the cultivars such as ICSV 700, MASV 33/93 and PB 15438 are tolerant to both the stem borer and the shoot fly (Singh and Shankar, 2000). Similarly there are several germplasm lines found to show multiple resistances to shoot fly and stem borer viz, KC-1, PGN-1, PGN-20 and PGN-64 (Kishore et al., 2002). The genotypes showing resistance to stemborer leaf feeding, deadheart formation, stem tunnelling, and/or compensation in grain yield can be used for sorghum improvement.

Shoot Bug

The genotypes Kafir Suma and Dwarf Hegari, I 753, H 109, GIB, 3677B, and IS 1055 in sorghum are free from *P. maidis* infestation (Khan and Rao, 1956; Agarwal et al., 1978). IS 19349, IS 18657, and IS 18677 were stable in resistance across different stages of crop growth (Chandra Shekar, 1991; Singh and Rana 1992; Chandra Shekar et al., 1992, 1993a,b).

Midge

Midge being a serious problem in sorghum across the globe, the screening of germplasm for resistance has been undertaken worldwide especially in Africa, Argentina, Australia, and El Salvador, India, Myanmar, USA (Bowden and Neve, 1953; Johnson et al., 1973; Rossetto et al., 1975; Shyamsunder et al., 1975; Jotwani, 1978; Faris et al., 1979; Page, 1979; Wiseman et al., 1988; Henzell et al., 1994; Sharma et al., 2002). Midge resistance in sorghum is polygenic (Henzell and Hare, 1996) like in many other insect pests.

The well known source of midge resistance, DJ 6514 confers resistance through the mechanisms of antixenosis and antibiosis while many of the North American derivatives possess largely antixenosis especially for oviposistion (Sharma, 1985; Henzell et al., 1994; Sharma et al., 2002). The midge resistance present in Australian hybrids is mainly drawn from North American sources (Henzell et al., 1994). The are several genotypes found to possess high levels of resistance to midge attack, IS 3461, IS 9807, IS 10712, IS 18563, IS 19476, IS 21873, IS 21881, IS 22806, ICSV 197, and PM 15936- 2 (Sharma et al., 2002).

CULTURING AND SCREENING TECHNIQUES

Screening for insect resistance under natural conditions suffers from serious lacunae such as highly fluctuating pest pressure, uncertainty of occurrence of particular pest and occurrence of huge errors in data accrued from field. Therefore, screening under proper conditions is very important to identify a true and lasting resistant source. Culturing of pest species is an essential prerequesit to ensure uniform pest pressure. There have been several attempts made to rear the insect pests under laboratory conditions using artificial diet, natural plant host, etc. Though many workers have attempted to rear shoot fly in laboratories, pot cultures and artificial diets (Soto and Laxminarayana, 1971; Dang et al., 1971; Sukhani and Jotwani, 1979; Singh et al., 1983), no satisfactory method has been developed.

Shoot fly, *A. soccata* has a unique life-history with females strictly choosing only the live host plant for oviposition. The maggots, mainly of second and third instars can be reared on sorghum (leaf and/or shoot) based decomposed material. Maggot does not necessarily need a specialized diet, unlike *C. partellus* or any other lepidopteran such as *Helicoverpa armigera* to grow in to pupa and adult under laboratory condition.

Several diets have been developed for mass rearing of *C. partellus* (Siddiqui et al., 1977; Taneja and Leuschner, 1985). The first artificial diet used to rear *C. parrellus* included many ingredients like leaf materials, salt mixture, yeast, casein, glucose, methyl paraben, cholestrol, choline chloride, cellulose, water and agar (Pant et al., 1960). The most popular chick-pea based diet has been used extensively to rear *C. partellus* (Dang et al., 1970).

Screening Techniques

Ensuring uniform level of insect pest pressure is an essential adjunct of effective screening programme. This can be achieved by testing plant material under artificial infestation with laboratory reared insects. Many a times testing the cultivars under hot-spot conditions are resorted to for screening large in large scale. Screening at a hot-spot location requires basic knowledge of insect population dynamics so that planting time can be adjusted to ensure that the susceptible stage of the crop coincides with the peak activity period of the insect.

There are methods developed such as infester row, artificial infestation, no-choice cage, detached leaf assay, and diet impregnation assay have been standardized to screen for resistance to shoot fly, stem borer, midge, and head bug in sorghum.

A field screening interlard-fishmeal technique using infester rows of susceptible cultivar and fishmeal baits provided sufficient and uniform infestation for large-scale testing of germplasms and breeding lines of sorghum (Soto, 1974). It is a simple technique in which fishmeal is moistened and kept in polyethylene bags. These bags with fishmeal is kept open and placed in the sorghum field. The shoot flies are attracted to the cues emanating from the decomposing fishmeal (Reddy et al., 1981). Several other flies are also attracted to the fishmeal apart from shoot flies. This technique to certain extend ensures enough shoot fly population pressure. Shoot flies can also be collected from fish-meal traps in the field and can be used for caged plant tests (Soto and Laxminarayan, 1971; Soto, 1972; Sharma et al., 1992; Dhillon et al., 2005).

In field screening it is not possible to ensure uniform initial infestation of all the plants in a population. Artificial infestation ensures a uniform and sufficient level of pest infestation at the desired time. Screening under artificial infestation is essential to confirm resistance observed under natural pest infestation as well as to study the mechanisms of resistance (Sharma et al., 1992; Dhillon et al., 2005). A three-step screening methodology was adopted for stem borer resistance testing in the AICRP on Sorghum (Pradhan et al., 1971). The first step was a general screening carried out in single-row plots under natural infestation. Selected materials were then entered into multirow replicated trials under natural infestation. The third step was confirmation of resistance in replicated trials under artificial infestation. In the on Sorghum programme, laboratory-reared insects have been released either as first-instar larvae (Singh et al., 1983) or as blackhead egg masses in the leaf whorls (Jotwani, 1978). For artificial infestation of *C. partellus*, insects are reared in the laboratory using an artificial diet, and for field infestation the *bazooka* applicator developed by Mihm and colleagues at the Centro International de Mejoramiento de Maiz y Trigo in 1976 (CIMMYT, 1977) can be used Sharma et al. (1992) and Dhillon et al. (2005).

Techniques to screen for midge resistance are available and a head cage technique for monitoring sorghum midge populations was developed at the ICRISAT Asia Centre, India. (Jotwani, 1978; Page, 1979; Sharma and Davies, 1988; Sharma et al., 1992). This technique is effective and efficient in collecting adult midges from flowering sorghum panicles under field conditions (Sharma and Davies, 1988; Sharma et al., 1992). Sorghum midge activity is higher during the rainy than during the post-rainy season in India (Kausalya et al., 1997). Screening at hot-spot locations is an important approach for identifying midge resistant genotypes. In India, the hot-spot locations for sorghum midge are Dharwad, Bhavanisagar and Pantnagar. In Africa, Sotuba in Mali, Farako Ba in Burkina Faso, Alupe in Kenya, and Kano in Nigeria are important hot-spot locations for sorghum midge (Sharma and Davies, 1988; Sharma et al., 1992).

BREEDING FOR RESISTANCE

Development of a successful resistant variety or hybrid depends on the following factors (Sharma et al., 1992; Dhillon et al., 2005; Tao et al., 2003): (1) a reliable screening technique, (2) developing criteria for measuring resistance, (3) identification of stable sources of resistance, (4) knowledge of the inheritance of resistance andthe mechanism of resistance, and (5) selection of breeding procedures to incorporate resistance into agronomically superior backgrounds.

Generally, resistance to many of the insect pests is quantitatively inherited and therefore difficult to transfer into high-yielding cultivars (Tao et al., 2003). For example, resistance to shoot fly is quantitatively inherited and polygenically controlled (Goud et al., 1983; Halalli et al., 1983; Agrawal and Abraham, 1985). Shoot fly resistance is due to gradual accumulation of genes and both additive and non-additive gene actions are involved, and a high degree of environmental influence govenrs the heritability of resistance (Sharma et al., 1977; Borikar and Chopde, 1980, 1981; Halalli et al., 1983; Nimbalkar and Bapat, 1992). Therefore, Genotype and Environmental (G x E) interactions play major role in developing pest resistant crop varieties, especially for shoot fly as its population varies with seasons and locations (Singh et al., 1978; Borikar and Chopde, 1982; Rana et al., 1984).

The additive gene action revealed by the estimates of the general combining ability (GCA) and specific combining ability (SCA) governs the inheritance for oviposition nonpreference, deadhearts, recovery resistance (tillering), and the morphological plant-traits associated with resistance or susceptibility to *A. soccata* (Dhillon et al., 2006). Similarly, estimates of SCA and heterosis indicate that heterosis breeding has no value in breeding for resistance to shoot fly (Dhillon et al., 2006).

A number of lines with resistance to shoot fly have been identified in the germplasm (Jotwani, 1978; Rana et al., 1981; Taneja and Leuschner, 1985; Sharma et al., 2003). Improved varieties such as CSV 5, CSV 6, CSV 7R, Swati (SPV 504), and CSV-8R developed using resistant land races are moderately resistant to shoot fly (Singh and Rana, 1986). However, the break through achieved in higher yield levels with releases of several high yielding cultivars of sorghum, have not been able to bring in a healthy resistance to shoot fly. This was one of the constraints in popularizing and utilization of improved hybrids and varieties and hence the average productivity is very low in India. In futherence of improving resistant levels for shoot fly, some of the improved lines such as ICSV 700, ICSV 705, and ICSV 717 have been developed in with shhot fly resistance and better yield potential (Agrawal and Abraham, 1985). In Africa, Serena and Seredo have been developed with high levels of recovery resistance (Dogget et al., 1970). A series of IS lines with high degree of stable resistance also identified, such as IS 1082, IS 2146, IS 5490, IS 5604, IS 4664, IS 1071, IS 2394, IS 5484 and IS 18368, IS 2146, IS 5566, IS 5469, IS 5490, and IS 1054 (Rao et al., 1977; Singh et al., 1978; Borikar and Chopde, 1982; Chundurwar et al., 1992).

Breeding for stem borer resistance started in 1966 in India, when a number of resistant parents were included in the breeding program (Pradhan et al., 1971). Since then a number of identified sources of resistance have been utilized by crossing them mostly with agronomically elite susceptible parents.

Several derivatives with high degrees of resistance have emerged from the crosses between a borer-resistant parent, BP53, and IS 2954. Through pedigree selection, a stem borer resistant pure line sorghum variety, P 311 is an example of resistant breeding programme using improved genotypes. There have been several such cultivars developed with resistance to shootfly, and stem borer such as SPV 1015 (PGS-1) DS-1, DS-2, DS-3, DS-4, DS-5 and DS-6 (Kishore, 1987, 1992, 1994).

Likewise breeding for other important pests have also been taken momentum and yielded several cultivars.

Resistance to *Melanaphis sacchari* was conditioned by a single major gene and was incompletely dominant (Tan et al., 1985). High levels of midge immunity have been incorporated from Indian,

American, and Australian breeding lines into elite, high-yielding sorghum varieties in India, Australia, and United States (Johnson et al., 1973; Henzell et al., 1980, 1994; Sharma et al., 1994; Jordan et al., 1998; Tao et al., 2003). Midge-resistant varieties ICSV 735, ICSV 758, and ICSV 804 have been released in Myanmar. Gene action for resistance to midge is largely governed by additive gene action. Resistance is needed in both the parents to produce midge-resistant hybrids (Sharma et al., 1994).

BIOTECHNOLOGICAL INTERVENTIONS

Quantitative Trait Locus Mapping: Marker-Assisted Selection

Quantitative trait locus (QTL) analysis is a statistical method to explain the genetic basis of complex traits using phenotypic data and genotypic data (Miles and Wayne, 2008). QTL analysis is used to link complex phenotypes to specific regions of chromosomes where the particular genetic locus responsible for a specific trait of interest is located.

Molecular markers linked to the specific traits of interest are used to map the QTL. Such molecular markers do not normally affect the specific trait of interest and segregate with the traits more frequently than the unlinked markers. Parents with genetically contrasting phenotypic expression for a specific trait of interest are selected for creating mapping populations. The heterozygous F_1 individuals are then crossed among themselves or backcrossed with parents to create mapping populations. These populations are scored for genotypic as well as phenotypic data.

The application of molecular markers for quantitative trait loci (QTL) analysis has provided an effective approach to dissect complicated quantitative traits into component loci to study their relative effects on specific trait (Doerge, 2002).

QTL analysis has been in use over several diverse fields and there has been several attempts made in crop improvement, especially for breeding insect resistant cultivars. There are some significant progress made in understanding the genetic basis of resistance and identification of QTLs by using molecular markers in sorghum for shoot fly, stem borer, aphids, shoot bug, head bug and midge (Ratnadass *et al.*, 2002; Aladele and Ezeaku, 2003; Tao *et al.*, 2003; Satish et al., 2009, 2012; Vinayan *et al.*, 2011; Wang FaMing *et al.*, 2013).

Shoot fly

Many of the plant traits reported to be associated with resistance to shoot fly in sorghum (refer table 5.1) are used as phenotypic characters for QTL mapping. The traits such as leaf glossiness, seedling vigour, seedling height, grain yield (Sajjanar, 2002), female oviposition preference, deadhearts, adaxial trichome density and abaxial trichome density (Satish *et al.*, 2012) have been mapped for QTL in sorghum. Instead of F2 populations, use of advanced generations, like recombinant inbred lines (RILs) is suggested to get unlimited amount of genetic material for repeated phenotyping. The QTL mapping data derived from RIL population can also be used to examine various methods of assessing quantitative genetic variation. Identified QTLs can be of great help in marker assisted selection (MAS) for conventional breeding for shoot fly resistance.

In sorghum, crosses between shoot fly resistant and susceptible lines are made to produce RIL populations. For example, RILs from cross between BTx63 (susceptible) and IS 18551 (resistant) used to identify QTLs for different component traits viz.,one for glossiness, two for seedling vigour, four for seedling height and one for grain yield (Sjjanar, 2002). In a cross between 296B (susceptible) and IS 18551 to identify genetic basis of shoot fly resistance in term of QTL (Satish *et al.*, 2009).

In these RIL populations, multiple QTL mapping (MQM) produced four each for leaf glossiness and seedling vigour, seven for oviposition, six for deadhearts, two for adaxial trichome density and six for abaxial trichome density. Both the resistant IS18551 the susceptible 296B parents contributed alleles for resistant with the share of the former being significantly large. QTL of the related component traits were co-localized, suggesting pleiotropy or tight linkage of genes. There is similarity in QTLs between sorghum and maize for insect resistance indicating genomic conservation of insect resistance loci in these crops. The genomic regions that harbor insect resistance QTLs can be utilized in introgression of resistance alleles into the susceptible high yielding background by marker-assisted backcross breeding. These genomic regions are helpful for enhancing the efficiency of selection and shortening the course of breeding cycle to screen target genotypes directly for related traits (Aruna *et al.*, 2011). The genes linked with specific traits of interest (insect resistance) in sorghum against shoot fly are Cysteine protease-Mir1 (XnhsbmSFC34/ SBI-10), significantly associated with major QTL for leaf glossiness, oviposition, deadhearts, adaxial trichome density and abaxial trichome density; NBS– LRR gene (XnhsbmSFCILP2/SBI-10), (in rice, this gene is associated withbrown planthopper resistance) is associated with deadhearts and oviposition; and beta-1,3-glucanase (XnhsbmSFC4/SBI-10), (associated with aphid and brown planthopper resistance) is linked to deadheart and leaf glossiness (Satish et al., 2012). There are common QTLs exist for shoot fly and other important sorghum insect pests such as greenbug, head bug, and midge. Markers associated with insect resistance QTLs can be used in breeding programmes to accelerate the process of developing stable resistant cultivars for shoot fly.

Stem borer

QTLs for five traits related to spotted stem borer resistanceviz., deadhearts, stem tunneling, leaf feeding damage, and recovery resistance are identified using RILs derived from a cross between ICSV 745 (susceptible) and PB 15520 (resistant). An important genomic region is located between *Xisep0829* and *XSbAGB02markers* on chromosome SBI-07 (Vinayan *et al.*, 2011).

Aphids

F_2 population derived from a cross between sorghum lines HN-16 and QS produced QTLs linked with microsatellite marker indicating the involvement of a single dominant gene in sorghum resistant to *Melanaphis sacchari* (Chang JinHua *et al.*, 2006). A cross between grain sorghum cultivar HN 16 (resistant) and BTx623 (susceptible) produced genetic populations segregating for *RMES1* (a resistant gene found in HN 16).

The closest markers flanking the *RMES1* locus are *Sb6m2650* and *Sb6rj2776*, which delimited a chromosomal region containing five predicted genes conferring resistance to *M. sacchari* (Wang FaMing *et al.*, 2013).

Midge

QTLs associated with two of the mechanisms of midge resistance, antixenosis and antibiosis are located on two different linkage groups. RILs derived from a cross between sorghum lines ICSV745 x 90562 showed QTLs associated with antixenosis trait, egg numbers/spikelet and with antibiosis explaining difference in egg and pupal counts. The identification of genes for different mechanisms of midge resistance will be particularly useful in exploring new sources of midge resistance and for gene pyramiding and breeding through MAS (Tao *et al.*, 2003).

Bacillus thuringiensis *Transgenics*

The biological insecticide from gram-positive soil bacterium, *Bacillus thuringiensis* (Bt) that produces a variety of crystalline proteins (d-endotoxins), popularly known as 'Cry toxins', or 'Cry proteins' during its sporulation phase of growth, are advantageous over chemical insecticides on several counts (Schnepf *et al.*, 1998; Kumar, 2003). With the advent of genetic transformation techniques, it has become possible to clone and insert genes into the crop plants to confer resistance to insect pests. Resistance to insects has been demonstrated in transgenic plants expressing genes for d-endotoxins from *B. thuringiensis*. Such transformed plants with foreign genes are called genetically modified crops or GM crops. Molecular biology and genetic engineering tools can be deployed to integrate naturally available insecticidal proteins in crops for various utilities (Ranjekar *et al.*, 2003).

Cry proteins have a narrow and specific spectrum of action against different pests, including dipteran and lepidopteran pests (Kumar *et al.*, 1996; Bates *et al.*, 2005; Bravo *et al.*, 2007; Van Frankenhuyzen, 2009). Various insecticidal crystal proteins (ICPs), or δ-endotoxins produced by *B. thuringiensis* (*Bt*) are very effective against lepidopteran insects (Hofte and Whiteley, 1989). CryIA CryIC, CryIE and CryIIA are effective against the larvae of spotted stem borer, *Chilo partellus* (Sharma *et al.*, 1999; Sharma et al., 2000).

Production of GM crops especially of *Bt* transgenics, in millets have been in a nascent stage mainly due to, the status of the crops being marginal and a general discouragement for GM crops in several countries. In sorghum, difficulties in tissue culture, regeneration, and genetic transformation (Emani *et al.*, 2002), could be an additional reason for slow pace of the application of transgenic approaches to genetic improvement of sorghum compared to other cereal crops (Zhong *et al.*, 1998). The first production of transgenic sorghum plants was by *Agrobacterium*-mediated transformation using immature embryos (Zhao *et al.*, 2000). Some of the earliest attempts on insect resistance in sorghum by introducing *cry1Ab* and *cry1B* genes into the sorghum genotype P898012 (Gray *et al.*, 2004), and a synthetic Bt cry1Ac gene (Girijashankar *et al.*, 2005) were made against stem borer, *C. partellus*. These attempts although achieved significant levels of reductions in leaf damage, larval mortality and larval weight, could not produce significant larval mortality and larval tunneling into young shoots. Further, the levels of Bt proteins produced were far below the lethal dose required to give complete protection against neonate larvae of *C. partellus* (Girijashankar *et al.*, 2005).

Transgenic sorghum plants were produced using synthetic *Bt* genes *(ubi1-cry1B)* by incorporating insect resistance into Indian lines via particle bombardment and *Agrobacterium*-mediated transformation methods. These transformed plants were resistant to *C. partellus* in terms of less leaf feeding and stem tunnelling (Balakrishna *et al.*, 2010; Visarada *et al.*, 2007).

Sorghum varieties 115, ICS21B and 5-27 transformed with cry1Ab gene with a relatively high level of Bt gene expression displayed insect-resistance to pink borer, *Sesamina inferens* (Zhang *et al.*, 2009). Transgenic sweet sorghum plants with cry1Ah showed high insect-resistance to *Ostrinia furnacalis* (Zhu *et al.*, 2011).

BIORATIONAL APPROACHES
PEST-EVASION TECHNIQUES

We intend pest-evasion techniques to mean the methods followed to avoid or divert pests from the millet crops. These techniques are agronomic or cultural methods.

Tillage

Deep ploughing during summer, also called summer ploughing, is a general practice recommended for all the crops in India. Such off-seasonal ploughing would expose the insect stages hiding underground to birds and other natural enemies.

Clean Cultivation

The bunds and fallow areas should be maintained weed free to prevent pest buildup on wild hosts. Shoot flies are known to survive and build up their population on wild grasses.

Adjustment of Sowing Dates

In India, early sowing of sorghum during the rainy season and late sowing during the postrainy season save the crop from shoot fly attack to some extent. In Africa, early sown pearl millet escapes heavy stem borer incidence, but coincides with peak activity of the earhead miner moth. Midge damage occurs more on late sown pearl millet. The date of sowing needs to be adjusted depending upon the pest occurrence in an area to escape pest attack.

Intercropping

Intercropping of millets with legumes is a common practice in India. This practice helps in reducing the pest pressure on both crops. Sorghum plus pigeon pea (2:1) intercropping coupled with early sowing and endosulfan spray significantly reduced the infestation of stem borer (Shekharappa and Kulkarni, 2003).

Crop Rotation

Crop rotation with noncereal crops is a healthy practice on several accounts such as maintenance of soil health and managing weeds, pests, and diseases. It is proven in several cropping systems, such as the rice–pulse cropping sequence in Asia. However, crop rotation with similar kinds of crops might pose some problems. Proso millet rotated with winter wheat in the United States effectively checks the proliferation of summer weeds while encouraging grasshoppers, mites, and wheat stem maggots to shift from wheat to proso millet and cause damage to the latter. Of late, in India too, there are some coastal deltaic regions like the Krishna River delta, where sorghum is cultivated after rice as rice–fallow sorghum. The pest and soil health issues need to be carefully monitored in such areas for any initial clues that could forecast any foreseeable larger trouble.

Trap Crops

The crops that have ready acceptance by insects for feeding can be used as trap crops. Trap crops are not generally tried with millets; however, castor bean and other potential trap crops can be used for trapping panicle-feeding bugs of sorghum (Ratnadass et al., 2011).

Companion Plants

An intercrop that influences the first trophic level by enhancing nutrition, or chemical defense through repelling, or intercepting effects on pests may be called a companion crop. Companion plants can attract natural enemies or provide food for natural enemies (Parolin et al., 2012). Certain aromatic plants such as marigold planted with main crops reduce the number of eggs laid by cabbage root fly and onion fly, both belonging to the anthomyiid genus *Delia*, and by lygaeus bugs (Finch et al., 2003). Such companion crops can be tried with millets for shoot fly. However, there are some conflicting reports on the role of companion crops in significantly reducing pests and increasing natural enemies (Moreno and Racelis, 2015).

Insectary Plants

A plant can attract natural enemies or provide food for natural enemies for an extended period so as to augment the natural enemies in a cropping ecosystem (Parolin et al., 2012). Flowering plants that can provide nectar to the adult natural enemies, usually perennials, are preferred as insectary plants.

BIOLOGICAL PEST SUPPRESSION

Biological pest suppression or biocontrol of insect pests has been successful in several agricultural ecosystems. The proponents of biological control of insect pests advocate three principal methods of biological pest suppression, namely inundation, inoculation, and augmentation. Inundation refers to the release of mass cultured biological agents, generally imported organisms. This is also called classical biological control or importation. Inoculation is the initial release of an organism in an area so as to enable it to further multiply and perpetuate in that ecosystem. Augmentation is the supplemental release of a cultured population of a species of bioagent that is already present in an ecosystem so as to boost its number. In the millets, there are very few artificial interventions for biological control (Bhatnagar, 1987). However, the millet crop ecosystems offer great opportunity for biological means of pest suppression as is evident from the presence of an astonishingly huge number of natural enemies reported for just one insect pest, the sorghum shoot fly (Singh and Sharma, 2002). The number grows even further when the natural enemy records of other pests such as stem borers, aphids, bugs, and midges are added.

Shoot Flies

A number of natural enemies, including spiders, have been recorded on eggs, larvae, and pupae of the sorghum shoot fly. Egg parasitoids of sorghum shoot fly are *Trichogrammatoidea bactrae*, *Trichogramma chilonis*, *Trichogramma evanescens*, *Trichogramma japonica*, *Trichogramma kalkae*, and *Trichogrammatoidea simmondsi* from India, Africa, and Europe (Deeming, 1971; Breniere, 1972; Taley and Thakare, 1979; Del Bene, 1986; Delobel and Lubega, 1984). *T. chilonis* and *T. simmondsi* are important mortality factors of *A. soccata* in India (Singh and Sharma, 2002; Kalaisekar et al., 2013). The egg parasitoids are extremely important in the management of shoot fly, because damage to the crop is a highly probable event once the larva enters the shoot. There are several species of parasitoids that attack shoot fly larvae. But the impact in terms of saving the crop from pest damage is almost nil, although it does have an influence on the population dynamics of the shoot fly. The important larval parasitoids are *Neotrichoporoides nyemitawus*, *Bracon* sp., and *Hockeria* sp. in India and Burkina Faso. Pupal parasitoids include *Spalangia endius*, *Trichopria* sp., *Opius* sp., *Monelta* sp. and *Rhoptromeris* sp., *Alysia* sp., *Trichoplasta* sp., and *Crataepiella* sp. from India, Africa, and Europe. The pupal parasitoid, especially *S. endius*, is an important population regulating parasitization of shoot flies in India (Kalaisekar et al., 2013). The most abundant parasitoids during the postrainy season in sorghum ecosystems are *Aprostocetus* sp., *N. nyemitawus*, *Opius* sp., and *S. endius*. Mass rearing techniques are available only for *T. chilonis*. Several species of predators are also active in the millet ecosystem, such as coccinellids, formicids, thrips, and spiders. Information on the diversity of pathogens of shoot fly populations is limited. A fungus, *Fusarium* sp., and a bacterium, *Corynebacterium* sp., have been isolated from shoot fly eggs (Zongo et al., 1993).

Stem Borers

A large number of natural enemies of stem borers have been identified (Table 5.3) for their potential use in biological control programs. Braconid species are the most abundant parasitoids recorded on the

Table 5.3 Parasitoids Recorded on the Sorghum Stem Borer *Chilo partellus* Swinhoe

Stage	Parasitoid
Egg	*Trichogramma chilonis*
Larva	*Macrocentrus sesamivorus*
	Cotesia sesamiae, Cotesia flavipes
	Chelonus curvimaculatus
	Goniozus indicus
	Bassus sp., *Chelonus* sp., *Cotesia* sp., *C. sesamiae*, *Cotesia ruficrus*, *Dolichogenidea* sp., *Dolichogenidea aethiopica, Dolichogenidea polaszeki, Megaselia* sp.
Pupa	*Pediobius furvus*
	Dentichasmias busseolae
	D. busseolae, *Brachymeria* sp., *Brachymeria olethria*, *P. furvus, Psilochalcis soudanensis, Syzeuctus ruberrimus*

sorghum stem borer *C. partellus*. Some classical biological controls have been tried against the borers in Africa and the United States. The larval parasitoid *Cotesia flavipes* was introduced into Kenya from Pakistan for biological control of *C. partellus*. *C. flavipes* is by far the most widely introduced parasitoid and in more than 40 countries it was introduced to control crambid stem borers (Polaszek and Walker, 1991). It has exemplarily high host-searching ability and is an important factor for the success of *C. flavipes*. The introduction of *Trichogramma ostriniae* from China to North America for the control of *Ostrinia nubilalis* was also a successful classical biological control. However, classical biological control of *C. partellus* in South Africa by the introduction of egg, larval, and pupal parasitoids yielded poor results because of climatic factors (Kfir, 1994). The most abundant and widespread parasitoids in the east African region are the egg parasitoids, *Telenomus* spp. and *Trichogramma* spp.; the larval parasitoids, *Cotesia sesamiae* and *Sturmiopsis parasitica*; and the pupal parasitoids, *Pediobius furvus* and *Dentichasmias busseolae*.

The larval parasitoids *Apanteles sesamiae*, *Tetrastichus atriclavus*, and *P. furvus* are effective in controlling *Busseola fusca* in Africa (Harris, 1962). The ichneumonid parasitoid *Syzeuctus africanus* and the bethylid *Goniozus procerae* parasitize the larvae of *Coniesta ignefusalis*.

Plant Hoppers

Several natural enemies of *P. maidis* are available in the field. The eggs of *P. maidis* are attacked by a wide spectrum of parasitoids and predators. Notable egg parasitoids are *Anagrus optabilis*, *Anagrus flaveolus*, *Anagrus osborni*, *Anagrus frequens* (Mymaridae: Hymenoptera), and *Ootetrastichus beatis* (Eulophidae: Hymenoptera). The adult parasitoid *Haplogonatopus vitiensis* belonging to Dryinidae (Hymenoptera) also occurs on *P. maidis*. Mirid bugs such as *Tytthus mundulus* and *Cyrtorhinus lividipennis* are predaceous on eggs and nymphs of *P. maidis*. *C. lividipennis* is considered an important density-dependent factor on adults and nymphs of *P. maidis*. The tettigonids *Conocephalus saltator* and *Xiphidiopsis lita* are predatory on early stages of the shoot bug. Coccinellid predators such as *Cheilomenes sexmaculata*, *Coccinella septempunctata*, and *Coelophora inaequalis* are found feeding on the young nymphs of *P. maidis* (Napompeth, 1973). The chrysopid *Mallada basalis* and syrphid *Allograpta exotica* are frequent predators on the nymphs of *P. maidis* (Beardsley, 1971; Napompeth, 1973). Spiders, *Plexippus paykulli*, *Hasarius adansoni*, *Pagiopalus atomarius*, *Argiope avara*, and *Tetragnatha mandibulata*, prey on the nymphs and adults (Napompeth, 1973). The red mite, *Brochartia* sp., also parasitizes both the nymphs and the adults of macropterous and brachypterous forms of the shoot bug (Kumar et al., 2005). A predatory ant, *Monomorium fossulatum*, was found feeding on eggs and early instar nymphs of the shoot bug (Kumar and Prabhuraj, 2005).

Aphids

The most abundant and extremely active natural enemy populations are associated with aphids in the millet ecosystem. There are over 47 species of natural enemies that attack the aphid *M. sacchari* world-wide and such a huge population, at times, keeps the aphids below the economic threshold levels in sorghum (van Rensburg, 1973; Meksongsee and Chawanapong, 1985; Singh et al., 2004). However, there are some factors such as cannibalism, intraguild predation, and interspecific competition that affect the predatory efficiency of natural enemies in the sorghum ecosystem (Kalaisekar et al., 2010). Lady beetles, lacewings, and hoverflies cause the greatest mortality to *M. sacchari* populations in the field (Singh et al., 2004). The important predators are the lady beetles, such as *Brumus suturalis, Cheilomenes propinqua, Cheilomenes lunata, Cheilomenes sulphurea, Chilocorus nigritus*, and *Scymnus morelleti*. The common lacewings are *Chrysoperla carnea* and *M. basalis*. The hoverflies, namely, *Ischiodon scutellaris* and *A. exotica*, are commonly found feeding on aphids. The common parasitoids of *M. sacchari* are *Aphelinus maidis, Lioadalia flavomaculata, Lysiphlebus delhiensis* and *Lysiphlebus testaceipes*.

Midges

The common parasitoids belonging to the Eupelmidae occurring on *Stenodiplosis* are *Eupelmus popa, Eupelmus australiensis, Eupelmus varieolor*, and *Eupelmus urozonus*; and those belonging to the Eulophidae are *Tetrastichus blastophagi, Aprostocetus diplosidis, Aprostocetus venustus, Ceratoneura petiolata*, and *Pediobius pyrogo* (Baxendale et al., 1983).

Millet Head Miners

In Africa, a series of successful augmentative mass releases of the gregarious larval ectoparasitoid *Habrobracon hebetor* (Hymenoptera: Braconidae) have been made to control the millet head miner, *Heliocheilus albipunctella*, starting in 1985 (Bhatnagar, 1987). *H. hebetor* is a larval parasitoid of several economically important lepidopteran pests. The method of release was to place jute bags containing pearl millet grain as well as flour along with parasitized host larvae in fields.

SEMIOCHEMICALS

Semiochemicals are chemicals involved in the biological communications between individuals of organisms. Semiochemicals are subdivided into allelochemicals and pheromones. Semiochemicals mediating interactions between interspecific individuals are called allelochemicals and that between intraspecific individuals are known as pheromones. Allelochemicals are chemicals that are significant to individuals of a species different from the source species. Allelochemicals are subdivided into allomones, kairomones and synomones. In case of allomones, the response of the receiver is adaptively favorable to the emitter but not the receiver; the kairomone is favorable to the receiver but not the emitter; and the synomone is favorable to both emitter and receiver. Pheromones (Gk. phereum, to carry; horman, to excite or stimulate) are released by one member of a species to cause a specific interaction with another member of the same species. Pheromones are highly species specific. Pheromones may be further classified as alarm, aggregation, and sex pheromone.

A deterrent pheromone is associated with the water-soluble glue with which the females of *Atherigona soccata* attach their eggs to the leaves (Raina, 1981). However, an oviposition stimulant present in the leaves of wild sorghum influences multiple egg laying by *A. soccata* (Salin and Kanaujia1999). Parasitoid cues could be interfering in the developmental biology of shoot fly, *A.soccata* (Kalaisekar, 2014).

The pheromonal communications in spotted stem borer, *Chilo partellus* has not yet been understood fully.

Pheromone baited traps are useful devices for monitoring moth population levels of stem borer. One of the earliest attempts on *C.partellus* pheromonal studies brought out two olfactory stimulants, (Z)-11-hexadecenal (I) and (Z)-11-hexadecen-1-ol (II) as the main components in the pheromonal compound that elicit significant responses in male moths (Nesbit *et al.,* 1979). The female sex pheromone components of *C. partellus* identified as (Z)-11-hexadecenal and (Z)-11-hexadecen-1-ol. As the age of the female moth progresses the release rate of the pheromone components reduced. The age-dependent shift in both release rate and ratio of pheromone components corresponds to the change in attractiveness of females to mate-searching males (Lux *et al.,* 1994; Gikonyo *et al.,* 2003).

Sex pheromone specific antennal sensilla of male *C. partellus* respond well to the combination of three pheromone components, (Z)-11-hexadecenal and (Z)-11-hexadecenol, and (Z)-10-pentadecenal (Hansson *et al.,* 1995).

Sex pheromones for *Chilo partellus* have been identified and are available commercially (Van den Berg and Nur, 1998) in Africa. Traps with female sex pheromone can be effectively used as monitoring tool for population fluctuations of *C. partellus.* However, under high population pressure, the pheromone catches might be misleading on the population levels. Positive correlations were observed between the number of males caught in pheromone traps and larval /pupal population density, males and percentage plants infested, and males and plant age. The flight phenology as monitored by pheromone traps appeared to reflect major events in population development of *C. partellus* (Unnithan and Saxena, 1990). The disruption in catch was not accompanied by a corresponding reduction in mating (Unnithan and Saxena, 1991). Trap catches influences significantly only the adult population in the field (Sharma, 1996).

Catches in pheromone traps indicate that the adults of *C. partellus* in fields of sorghum are most active between 23.00 and 03.00 h (Ho and Reddy, 1983).

The orientation of males of cecidomyiid *Stenodiplosis sorghicola* to sticky traps baited with virgin females in sorghum fields can induce a positive attraction response (Sharma and Vidyasagar, 1992).

BOTANICALS

Pesticides derived from plants have the potential to play a major role in eco-friendly pest management. They are renewable, non-persistent in the environment, and relatively safe to natural enemies, non-target organisms, and human beings.

The spray of neem extracts significant increases mortality in egg and larval of *Atherigona soccata* (Zongo *et al.,* 1993; Singh and Batra, 2001). Extracts of neem and custard apple kernels effectively reduces the damage by many pests such as *Chilo partellus, Mythimna separata, Calocoris angustatus* and *Melanaphis sacchari* in sorghum. Biologically active fraction of fresh neem kernels at 0.5% gives effective protection against *C. partellus, M. separata, Myllocerus* sp. and *C. angustatus.,* Piperine an extract of *Piper guineense* and its dihydrosaturated derivative are potent antifeedants for *C. partellus* and the presence of a methylenedioxybenzene and an alicyclic amide group in the compound may be crucial for high antifeedant activity (Torto *et al.,* 1992).

Powdered forms of different parts of *Tanacetum cinerariaefolium, Nicotiana tabacum, Azadirachta indica, Jatropha curcas, Cissus quadrangularis, Chenopodium ambrosioides* and *Euphorbia schimperiana* powders. *T. cinerariaefolium, N. tabacum, J. curcas* and *E. schimperiana* are potentially reducing damage of stemborer *C. partellus* on sorghum (Asmare Dejen *et al.,* 2011).

CHEMICAL CONTROL

In the millets, especially sorghum and pearl millet, productivity increased rapidly with the introduction of high-yielding cultivars during the 1970s and 1980s. As a result insect pests also made headway and acted as a major threatening factor for grain and fodder production systems. As an immediate remedy to manage the menace of pests, chemical pesticides are always relied upon mainly on high-yielding varieties and hybrids.

Chemical-based crop protection has economic feasibility only in farming sectors with commercial scale production such as seed production. Chemical management methods involve the use of insecticides to kill insect pests. They are powerful tools and practical control measures for insect pests. Insecticides have rapid, curative action but are costly and may cause negative ecological and environmental consequences. Therefore, their use must be judicious and based on actual measurement of insect abundance and damage. Insecticides should be applied only when the insect pest load is increasing and expected to exceed the economic injury level if not suppressed. There are various methods of insecticide application such as seed treatment, soil application, foliar application, and earhead application, depending upon the target pest.

The chemical control of insect pests in millets has not been practiced intensively, unlike in crops like rice, corn, and other cash crops. There have been several recommendations of chemical control in millets since the late 1950s. A brief history of chemical control in millets may be pertinent to know before looking at the present status.

HISTORY OF CHEMICAL CONTROL IN MILLETS

The historical background of chemical control in millets is discussed here to show the difficulties and disadvantages of various insecticides and their methods of application.

All the chemicals mentioned in this section are only to narrate the history of their use in the past and therefore, *chemicals mentioned here are strictly not recommended for use in millets.*

Shoot Fly Control

Shoot flies are one of the most important pests in millets, especially in sorghum (Sharma, 1993). Under favorable conditions an extent of damage of up to 90% has been reported (Chundurwar et al., 1992) and economic losses up to US$120 million (ICRISAT, 1992). The incidences of shoot fly increases with delayed sowings.

In India, before the introduction of high-yielding hybrids, cultural practices like early sowing, using increased seed rate, and pulling and destroying infested seedlings were recommended for reducing shoot fly damage. The first attempt made to control the pest with insecticide dusts and sprays of benzene hexachloride (BHC) and dichlorodiphenyltrichloroethane (DDT) was in India (Rao and Rao, 1956). Weekly applications of DDT–BHC dust were used for control of shoot flies in Tanzania (Swaine and Wyatt, 1954). In Uganda, an increase in shoot fly infestation was observed after the application of DDT and carbaryl (Davies and Jowett, 1966). DDT, carbaryl, and endrin sprays were ineffective against shoot flies in India (Vedamoorthy et al., 1965). Three sprays of 0.05% phosalone, diazinon, or demeton-*S*-methyl (Meta-Systox) were as effective as 3% granules of carbofuran applied in the soil. Cypermethrin spray was effective in the reduction of ovipositioning, resulting in less damage, due to either an ovipositioning deterrent effect or the death of the adults before ovipositioning (Taneja and Henry, 1993). In general, the foliar sprays were largely ineffective in controlling the fly (Boonsom et al., 1970) because many times, the larva of the shoot fly enters into the plant shoot and escapes by not having a chance to contact the insecticide applied on the foliar surface; in addition, the larva does not feed on foliage.

The application of systemic insecticides in soil either in powder or in granular form was a practice followed at sowing in some parts of India. Systemic granular insecticides like phorate, disulfan, and carbofuran applied in seed furrows at the time of sowing controlled shoot fly (Sandhu and Young, 1974). However, there were very few takers to use such methods of chemical application because of cost considerations (Sukhani and Jotwani, 1980). Another granular insecticide, disulfoton, was also used through soil application, recommended for effective control of shoot fly (Bhanot et al., 1984). Late sown crops in monsoon season and early sown crops in the postrainy season suffered most from the shoot fly, and under such conditions, soil application with carbofuran was recommended in India (Thimmaiah et al., 1973; Usman, 1973). However, the effectiveness of soil-applied granular insecticides is mainly dependent on soil moisture (Taneja and Henry, 1993). Soil application of granular insecticides under dry conditions, which is the most prevalent growing environment of millets, is therefore not of much use.

Treating seeds (seed dressing) is a more preferable and convenient method for growers to protect millets, especially sorghum, from shoot fly attack than treating through the soil application method. Seed treatment, which could protect the vulnerable seedling stage from shoot fly attack for about 30 days, is not only convenient to apply but also economical. Seed dressing with carbofuran provided the best results. The effectiveness of this method of insecticide application is also influenced by soil type and texture and better results are seen in light soils than in heavy ones (Jotwani and Young, 1972). Seed treatment with carbofuran was effective at 5 parts a.i. (active ingredient) of carbofuran per 100 parts of sorghum seed (Jotwani et al., 1971).

The introduction of hybrids in sorghum, namely CSH1 and CSH2, witnessed a steep increase in shoot fly incidences in India. Weights of ears and green fodder positively correlated with the degree of shoot fly control. Phorate 10G at 16 kg/acre as soil application or 6% carbofuran as seed treatment was recommended. Seed treatment was cost-effective (Balasubramanian et al., 1976). Several edaphic factors such as soil type, soil pH, cation exchange capacity, amount of organic matter, and soil phosphorus and potash contents affected the efficacy of chemicals applied as seed treatment as well as soil application.

Under high shoot fly population pressure, no single control measure is able to save the crop loss. The sole method of cultural control, early sowing in rainy season, is ineffective when shoot fly attack is considerably heavy. Under high shoot fly pressure even carbofuran seed treatment is rendered ineffective (Chaudhari et al., 1994).

Among all the kinds of chemical applications, seed treatment is the most effective chemical method for controlling shoot fly during the 30 days of seedling stage. However, from time to time there are changes in the chemical molecules and their formulation. Therefore, the efficacy of such new chemicals needs to be ascertained before suggesting chemical control measures for shoot fly.

Stem Borer Control

Lepidopteran stem-boring insects cause considerable crop loss throughout the millet-growing areas of the world. In India, the incidence of stem borers is recorded as from 10% to 75% with severe infestations (Pradhan and Prasad, 1955). Damage from 60% to 100% has been reported in various parts of India (All India Coordinated Sorghum Improvement Project, 1975–87).

Chemical control is an important means for managing stem borers. Chemical control had hardly been an option for borer control in east Africa until the late 1970s (Kayumbo, 1976) and the chemicals were ineffective in South Africa (van Rensburg and van Hamburg, 1975). In India, the earliest insecticides used for stem borer control were BHC and DDT, applied as sprays and dusts (AICSIP, 1975–87). Granular insecticides applied in whorls were effective but costlier. Insecticides such as endosulfan 4%, carbaryl 5%, lindane 0.65%, and phenthoate 2%, applied as dusts at reduced dosages of 8–10 kg/ha in the leaf whorls, were effective and economical (Jotwani, 1982). *C. partellus* was effectively controlled with dust formulations of BHC, carbaryl, malathion, endosulfan, and phenthoate (Natarajan and Chelliah, 1986). Spray formulations of parathion, diazinon, trichlorphos, carbaryl, malathion, and endosulfan were effective in reducing the pest (Sukhani, 1986).

The effectiveness of chemical pesticides against stem borers depends on factors such as the application method, formulation, and time of application.

Soil application: Furrow application of systemic insecticides such as cytrolane 5G, carbofuran 5-10G, aldicarb 10G, mephosfolan 10G, phorate 10G, and disulfoton 10G, at 0.7–2.0 kg a.i./ha.

Seed dressing: BHC and lindane BHC (5 g/100 g) and lindane (20 g/100 g) as seed treatment.

Side dressing: Mephosfolan and carbofuran at 1.0–2.5 kg a.i./ha, 20 days after emergence.

Foliar application: Foliar sprays and dusts of DDT, BHC, and parathion, and pyrethroids such as decamethrin, fenvalerate, cyloxylate, cypermethrin, and permethrin at 25–150 g a.i./ha.

Whorl application: Granules are directly placed into the whorl of sorghum and other millets. Carbofuran at 2.5 kg/acre 25 and 35 days after germination.

Granules: Granular application of BHC, endrin, carbofuran, chlordimeform, chlorfenvinphos, diazinon, fensulfothion, mephosfolam, or salithion at 0.45–2 kg a.i./ha.

Sprays: DDT, BHC, and parathion.

Dusts: DDT, endrin, endosulfan, phenothoate, carbaryl, malathion, and BHC.

Aphid Control

Aphid populations increase rapidly when the weather is dry or at the end of the rainy season. In sorghum, overall losses of 16% and 15% for grain yield and fodder yield, respectively, have occurred (Balikai, 2001). Aphids affect not only the grain and fodder yield but also the fodder quality. Incidences and yield losses are high in postrainy sorghum. Foliar and granular applications of systemic insecticides help reduce the aphid population.

Seed treatment: Thiamethoxam 70 WS at 3 g/kg seed against aphids in postrainy sorghum.

Soil application: Systemic insecticides such as disulfoton, disyston, and phorate.

Granular application: Whorl application of phorate, dithiodemeton, telodrin, and endosulfan.

Foliar application: Demeton-*S*-methyl, disulfoton, monocrotophos, parathion, dimethoate, malathion, demeton, carbophenothion, endrin, endosulfan, phorate, carbofuran, diazinon, Meta-Systox, phosdrin, quinalphos.

Application of chemicals for aphid control should be planned cautiously by considering the large complex of natural enemies. The intensity of infestation depends on aphid numbers, plant size, stage of growth, growing conditions, and activity of beneficial insects in grain or forage sorghums. These factors need to be considered before applying an insecticide for aphid management.

Shoot Bug Control

P. maidis was first recorded on corn in Hawaii by Perkins in 1892 (Zimmerman, 1948). It is difficult to accurately estimate the losses associated with specific levels of damage in terms of reduction in crop yields. In India, multiple losses were recognized and estimated at 10–15% loss from leaf sugar exudation (Mote et al., 1985; Mote and Shahane, 1993), 10–18% loss of plant stand (Managoli, 1973), and 30% loss of grain sorghum yield (Mote et al., 1985). Economic injury level is four nymphs per plant in sorghum. *P. maidis* is a serious menace to sorghum in soils of low fertility and in bunded areas (Borade et al., 1993).

The earliest chemicals recommended for its control in India were BHC, endrin, chlordane, isodrin, DDT, and aldrin dusts.

The chemicals used for foliar application were demeton-*S*-methyl, monocrotophos, carbaryl and phosphamidon, endosulfan malathion, and parathion.

Sorghum Midge Control

Two sprays at an interval of 4 days, starting 4 days after 90% of the heads have emerged from the boot, is the right timing for the effective control of the midge. Timing of application and choice of insecticide with no phytotoxicity and no residual effect are the important factors in midge control using chemicals.

For foliar application, DDT and BHC sprays were used for the control of the midge in Australia. In the United States, sprays of diazinon, carbaryl, dimethoate, endrin, methyl parathion, and toxaphene were recommended. Insecticides in a two-spray application schedule, 4 days apart, starting at 50% head emergence gave most effective control. In India, sprays of 0.07% endosulfan and 0.3% carbaryl and dusts containing 2% endosulfan, 10% carbaryl plus 10% HCH (BHC), and 5% malathion were recommended.

Chinch Bug (Blissus leucopterus)

Dieldrin, endrin, toxaphene, and endosulfan were used. Granular insecticides applied as a side dressing were effective.

Spider Mites (Oligonychus indicus)

Damage to sorghum crop by tetranychid mites has been reported from India, the United States, and Mexico. Sulfur dusting or spraying methyl demeton was recommended.

Earhead Pests

Earhead worms and bugs damage developing grains. Sprays of carbaryl, parathion, and phosphamidon were recommended.

PRESENT STATUS OF CHEMICAL CONTROL IN MILLETS

In India, around 7% of total pesticides consumed are used in cereals (excluding paddy), millets, and oilseeds. Table 5.4 provides the insecticides and their recommended dosages (registered under the Central Insecticide Board, India, as of August 10, 2015) for the control of pests of millets. The pesticide given under this heading are the ones that are in use at present and therefore, can be used against the specified insect pests. Further, several other insecticides, which are not mentioned here, are in use in India as well as elsewhere in the world. Such insecticides are not mentioned in this book due to lack of authentic evidential publications.

Shoot Fly

The most commonly used insecticide in millets is imidacloprid through seed treatment, generally for the control of shoot flies in India.

In sorghum, seed treatment with thiamethoxam 70 WS at 3 g/kg of seeds or imidacloprid 70 WS at 5 g/kg seeds is effective against shoot fly (Balikai, 2011).

In pearl millet, two sprays of profenofos 0.05% or fenobucarb 0.1% at 20 and 40 days after germination give effective control against shoot fly and stem borer (Parmar et al., 2015).

In little millet, seed treatment with imidacloprid 600FS at 5 mL/kg of seed can be recommended for shoot fly control (Kumar and Channaveerswami, 2015).

In kodo millet, seed treatment with thiamethoxam at 2 g/kg of seed and foliar spray of monocrotophos at 1.5 mL/L of water at 15 days after germination are ideal for shoot fly control.

The seed treatment with imidacloprid is costlier than with chloropyrifos and, therefore, the latter may be used if the cost is a constraint (Balikai, 2007).

Stem Borers

Two sprays of profenofos 0.05% or fenobucarb 0.1% at 20 and 40 days after germination control shoot fly and stem borer infestation of pearl millet (Parmar et al., 2015). Application of acetamiprid at transplanting and at 45 and 75 days after transplanting (DAT) or neem seed extract application at transplanting and at 30, 45, 60, 75, 90, and 105 DAT considerably reduced losses due to the stem borer *Sesamia cretica* in transplanted sorghum in Africa (Aboubakary et al., 2008). Sprays of aqueous neem seed extract, neem seed oil, and deltamethrin 12EC are equally effective for the management of sorghum whorl larvae, stem borers, and panicle insect pests in the Nigerian Sudan Savanna (Anaso, 2010).

Shoot Bugs

Seed treatment with thiamethoxam 70 WS at 2 g/kg, imidacloprid 70 WS at 5 g/kg, and carbosulfan 25 DS at 40 g/kg is recommended (Vijay Kumar and Prabhuraj, 2007).

Midges

Two sprays at an interval of 4 days, starting 4 days after 90% of the heads have emerged from the boot, is the right timing for the effective control of the midge.

PRECAUTIONS DURING CHEMICAL APPLICATION

Residues

A general waiting period of 40–60 days is safer for harvest or consumption.

Table 5.4 Insecticides for Pests of Millets (Registered Under Central Insecticide Board, India, as of August 10, 2015)

Insecticide	Active Ingredient (g/ha)	Formulation (g/ha)
Sorghum: Shoot fly		
Carbofuran 3% CG (soil application)	1000	33,300
Carbaryl 50% WP (spray)	750	1,500
Imidacloprid 48% FS (seed treatment)	7.2 g/kg	12 g/kg
Imidacloprid 70% WS (seed treatment)	7 g/kg	10 g/kg
Oxydemeton-methyl 25% EC (spray)	250	1,000
Phorate 10% CG (soil application)	1875	18,750
Quinalphos 25% EC (spray)	375	1,500
Thiamethoxam 30% FS (seed treatment)	3 g/kg	10 g/kg
Sorghum: Stem borer		
Carbofuran 3% CG (whorl application)	250	8,300
Carbaryl 50% WP (spray)	1000	2,000
Quinalphos 5% G (soil application)	750	15,000
Sorghum: Aphids		
Carbaryl 10% DP (spray)	2500	25,000
Carbaryl 50% WP (spray)	1000	2,000
Phorate 10% CG (soil application)	1875	18,750
Sorghum: Hoppers, bugs		
Carbaryl 50% WP (spray)	1000	2,000
Sorghum: White grub		
Phorate 10% CG (soil application)	2500	25,000
Sorghum: Mites		
Phenthoate 2% DP (spray)	400	20,000
Quinalphos 25% EC (spray)	375	1,500
Sorghum: Earhead midge		
Carbaryl 5% DP (spray)	1000	20,000
Dimethoate 30% EC (spray)	500	1,650
Malathion 5% DP (spray)	1000	20,000
Malathion 50% EC (spray)	500	1,000
Phosalone 35% EC (spray)	400	1,143
Phosalone 4% DP (spray)	1000	25,000
Quinalphos 1.5% DP (spray)	400	26,600
Sorghum: Earhead caterpillar		
Carbaryl 10% DP (spray)	2000	20,000

Continued

Table 5.4 Insecticides for Pests of Millets (Registered Under Central Insecticide Board, India, as of August 10, 2015)—cont'd

Insecticide	Active Ingredient (g/ha)	Formulation (g/ha)
Sorghum: Earhead bug		
Quinalphos 1.5% DP (spray)	375	25,000
Pearl millet: Shoot fly		
Carbofuran 3% CG (soil application)	1500	50,000
Imidacloprid 48% FS (seed treatment)	7.2 g/kg	12 g/kg
Imidacloprid 70% WS (seed treatment)	7 g/kg	10 g/kg
Pearl millet: Termite		
Imidacloprid 48% FS (seed treatment)	7.2 g/kg	12 g/kg
Imidacloprid 70% WS (seed treatment)	7 g/kg	10 g/kg
Pearl millet: Milky weed bug		
Dimethoate 30% EC (spray)	180–200	594–660

Phytotoxicity

Chemicals that are phytotoxic to millets should not be used.

Personal Care and Protection

All the protective measures such as wearing a mask and coat must be followed while applying insecticides in the field.

REFERENCES

Aboubakary, Ratnadass, A., Mathieu, B., 2008. Chemical and botanical protection of transplanted sorghum from stem borer (*Sesamia cretica*) damage in northern Cameroon. Journal of SAT Agricultural Research 6, 1–5.

Agarwal, R.K., Verma, R.S., Bharaj, G.S., 1978. Screening of sorghum lines for resistance against shoot bug, *Peregrinus maidis* Ashmead (Homoptera: Delphacidae). JNKVV Research Journal 12, 116.

Agrawal, B.L., Abraham, C.V., 1985. Breeding sorghum for resistance to shoot fly (*Atherigona soccata*) and midge (*Contarinia sorghicola*). In: Proceedings of the International Sorghum Entomology Workshop, 15–21 July 1984. International Crops Research Institute for the Semi-Arid Tropics, Patancheru, Andhra Pradesh, India, pp. 371–383.

Aladele, S.E., Ezeaku, I.E., 2003. Inheritance of resistance to head bug (*Eurystylus oldi*) in grain sorghum (Sorghum bicolor). African Journal of Biotechnology 2, 202–205.

Alghali, A.M., 1985. Insect-host plant relationships. The spotted stalk-borer, *Chilo partellus* (Swinhoe) (Lepidoptera: Pyralidae) and its principal host, sorghum. Insect Science and Its Application 6, 315–322.

AICSIP (All India Coordinated Sorghum Improvement Project), 1975–87. Progress Reports of the All India Coordinated Sorghum Improvement Project, Indian Council of Agricultural Research and Cooperative Agencies. AICSIP, New Delhi, India.

Anaso, C.E., 2010. Efficacy of neem pesticides on whorl larva, stem-borer and panicle insect pests of sorghum in Nigeria. Archives of Phytopathology and Plant Protection 43 (7/9), 856–862.

Aruna, C., Bhagwat, V.R., Madhusudhana, R., Sharma, V., Hussain, T., Ghorade, R.B., Khandalkar, H.G., Audilakshmi, S., Seetharama, N., 2011. Theoretical and Applied Genetics 122, 1617–1630.

Balakrishna, D., Padmaja, P.G., Seetharama, N., 2010. *Agrobacterium*-mediated transformation of sorghum with synthetic cry1B gene for resistance against spotted stemborer (*Chilo partellus*). In: National Symposium on Genomics and Crop Improvement: Relevance and Reservations at ANGRAU during 25–27, February, 2010.

Balasubramanian, R., Thontadarya, T.S., Heinrichs, E.A., 1976. Chemical control of the sorghum shoot fly *Atherigona varia* var. soccata Rondani (Diptera: Anthomyiidae) in South India. Mysore Journal of Agricultural Sciences 10 (2), 245–251.

Balikai, R.A., 2001. Bioecology and Management of the Sorghum Aphid, *Melanaphis sacchari* (Ph.D. thesis). University of Agricultural Sciences, Dharwad, Karnataka, India. 203 pp.

Balikai, 2007. Eco-biology and management of aphid *Melaphis sacchari* (Zehntner) in rabi sorghum. Agriculture Reviews 28 (1), 18–25.

Bates, S.L., Zhao, J.Z., Roush, R.T., Shelton, A.M., 2005. Insect resistance management in GM crops, past, present and future. Nature Biotechnology 23, 57–62.

Baxendale, F.P., Lippincott, C.L., Teetes, G.L., 1983. Biology and seasonal abundance of hymenopterous parasitoids of sorghum midge (Diptera: Cecidomyiidae). Environmental Entomology 12, 871–877.

Beardsley, J.W., 1971. Notes and exhibitions. Proceedings of the Hawaiian Entomological Society 21, 21.

Bergquist, R.R., Rotar, P., Mitchell, W.C., 1974. Midge and anthracnose head blight resistance in sorghum. Tropical Agriculture (Trinidad) 51, 431–435.

Bhagwat, V.R., Shyam Prasad, G., Kalaisekar, A., Subbarayudu, B., Hussain, T., Upadhyaya, S.N., Daware, D.G., Rote, R.G., Rajaram, V., 2011. Evaluation of some local sorghum checks to shoot fly, Atherigona soccata Rondani and Stem borer, *Chilo partellus* Swinhoe resistance. Annals of Arid Zone 50 (1), 47–52.

Bhanot, J.P., Verma, A.N., Lodhi, G.P., 1984. Control of sorghum shoot fly (*Atherigona soccata* Rondani) with systemic granular insecticides. Haryana Agricultural University Journal of Research 14 (1), 89–91.

Bhatnagar, V.S., 1987. Conservation and encouragement of natural enemies of insect pests in dry land subsistence farming: problems, progress and prospects in the Sahelian zone. Insect Science and Its Application 8, 791–795.

Blum, A., 1967. Varietal resistance in sorghum varieties resistant Bapat, D. R. and Mote, U. N. 1982a. Upgrading the resistance of derivatives from Indian × Indian crosses of sorghum against shoot fly. Journal of Maharashtra Agricultural Univerisites 7, 170–173.

Blum, A., 1968. Anatomical phenomena in seedlings of sorghum varieties resistant to the sorghum shoot fly, (*Atherigona varia soccata*). Crop Science 8, 388–391.

Blum, A., 1969. Factors associated with tiller survival in sorghum varieties resistant to the sorghum shoot fly (*Atherigona varia soccata*). Crop Science 9, 508–510.

Blum, A., 1972. Sorghum breeding for shoot fly resistance in Israel. In: Jotwani, M.G., Young, W.R. (Eds.), Control of Sorghum Shoot Fly. Oxford and IBH Publishing Co., New Delhi, pp. 180–191.

Boonsom, M., Knapp, F.W., Sepasawadi, P.R., Wongwiwat, Wongobrat, A., 1970. Report of the Thailand Corn and Sorghum Program Annual Reporting Session, Bangkok, Thailand, January 15–16, 1971, pp. 197–201.

Borade, B.V., Pokharkar, R.N., Salunkhe, G.N., Gandhale, D.N., 1993. Effects of dates of sowing on leaf sugar malady on rabi sorghum. Journal of Maharashtra Agricultural Universities 18, 124–125.

Borikar, S.T., Chopde, P.R., 1980. Inheritance of shoot fly resistance under three levels of infestation in sorghum. Maydica 25, 175–183.

Borikar, S.T., Chopde, P.R., 1981. Inheritance of shoot fly resistance in sorghum. Journal of Maharashtra Agricultural Universities 6, 47–48.

Borikar, S.T., Chopde, P.R., 1982. Genetics of resistance to sorghum shoot fly. Zeitschrift fur Pflanzenzuchtung 88, 220–224.

Bowden, J., Neve, R.A., 1953. Sorghum midge and resistant varieties in the Gold Coast. Nature 172, 551.

Bravo, A., Gill, S.S., Soberon, M., 2007. Mode of action of *Bacillus thuringiensis* Cry and Cyt toxins and their potential for insect control. Toxicon 49, 423–435.

Breniere, J., 1972. Sorghum shoot fly in west Africa. In: Jotwani, M.G., Young, W.R. (Eds.), Control of Sorghum Shoot Xy. Oxford & IBH Publishing Company, New Delhi, India, pp. 129–136.

Chandra Shekar, B.M., Dharma Reddy, K., Singh, B.U., Reddy, D.D.R., 1992. Components of resistance to corn planthopper, *Peregrinus maidis* (Ashmead), in sorghum. Resistance Pest Management Newsletter 4, 25.

Chandra Shekar, B.M., Reddy, K.D., Singh, B.U., Reddy, D.D.R., 1993a. Antixenosis component of resistance to corn planthopper, *Peregrinus maidis* (Ashmead) in sorghum. Insect Science and Its Application 14, 77–84.

Chandra Shekar, B.M., Singh, B.U., Reddy, K.D., Reddy, D.D.R., 1993b. Antibiosis component of resistance in sorghum to corn planthopper, *Peregrinus maidis* (Ashmead) (Homoptera: Delphacidae). Insect Science and Its Application 14, 559–569.

Chandra Shekar, B.M., 1991. Mechanisms of Resistance in Sorghum to Shoot Bug, *Peregrinus maidis* (Ashmead) (Homoptera: Delphacidae) (M.Sc. thesis). Andhra Pradesh Agricultural University, Hyderabad 500 030 (AP), India. 106 pp.

Chaudhari, S.D., Karanjkar, R.R., Chundurwar, R.D., 1994. Efficacy of some insecticides to control shoot fly, delphacid, aphids and leaf sugary disease in sorghum. Entomon 9, 63–66.

Chundurwar, R.D., Karanjkar, R.R., Borikar, S.T., 1992. Stability of shoot fly resistance in sorghum. Journal of Maharashtra Agricultural Universities 17, 380–383.

CIMMYT (Centro Internacional de Mejoramiento de Maiz y Trigo), 1977. CIMMYT Review 1977. CIMMYT. El Batan, Mexico, p. 99.

Dabrowski, Z.T., Kidiavai, E.L., 1983. Resistance of some sorghum lines to the spotted stalk borer, Chilo partellus, under western Kenyan conditions. Insect Science and Its Applications 4, 119–126.

Dahms, R.G., 1969. Theoretical Effects of Antibiosis on Insect Population Dynamics. United States Department of Agriculture, ERO, Beltsville, p. 5.

Dang, K., Anand, M., Jotwani, M.G., 1970. A simple improved diet for mass rearing of sorghum stem borer, Chilo zonellus (Swinhoe). Indian Journal of Entomology 32, 130–133.

Dang, K., Doharey, K.L., Srivastava, B.G., Jotwani, M.G., 1971. Artificial diet for the mass rearing of *Atherigona soccata* Rondani. Entomologists' Newsletter 1, 64–65.

Davies, J.C., Jowett, D., 1966. Increases in the incidence of *Atherigona indica infuscate* Emden (Diptera: Anthomyiidae) on sorghum due to spraying. Nature 209, 104.

Deeming, J.C., 1971. Some species of *Atherigona* Rondani (Diptera: Muscidae) from Northern Nigeria, with special reference to those injurious to cereal crops. Bulletin of Entomological Research 61, 133–190.

Dejen, A., Getu, E., Azerefegne, F., Ayalew, A., 2011. Efficacy of some botanicals on stem borers, *Busseola fusca* (Fuller) and *Chilo partellus* (Swinhoe) on sorghum in Ethiopia under field conditions. Biopesticides International 7, 24–34.

Del Bene, G., 1986. Note sur la biologia di *Atherigona soccata* Rondani (Diptera, Muscidae) inToscano e Lazio. Redia LXIX 47–63.

Delobel, A.G.L., Lubega, M.C., 1984. Rainfall as a mortality factor in the sorghum shoot fly, *Atherigona soccata* Rondani (Diptera: Muscidae) in Kenya. Insect Science and Its Application 2, 67–71.

Denno, R.F., Roderick, G.K., 1990. Population biology of planthoppers. Annual Review of Entomology 35, 489–520.

Deu, M., Ratnadass, A., Hamada, M.A., Noyer, J.L., Diabate, M., Chantereau, J., 2005. Quantitative trait loci for head-bug resistance in sorghum. African Journal of Biotechnology 4 (3), 247–250.

Dhawan, P.K., Singh, S.P., Verma, A.N., Arya, D.R., 1993. Antibiosis mechanism of resistance to shoot fly, *Atherigona soccata* (Rondani) in sorghum. Crop Research 6, 306–310.

Dhillon, M.K., Sharma, H.C., Singh, R., Naresh, J.S., 2005. Mechanisms of resistance to shoot fly, *Atherigona soccata* in sorghum. Euphytica 144 (3), 301–312.

Dhillon, M.K., Sharma, H.C., Reddy, B.V.S., Singh, R., Naresh, J.S., 2006. Inheritance of resistance to sorghum shoot fly, *Atherigona soccata*. Crop Science 46 (3), 1377–1383.

Dhillon, M.K., 2004. Effects of Cytoplasmic Male-Sterility on Expression of Resistance to Sorghum Shoot Fly, *Atherigona soccata* (Rondani) (Ph.D. thesis). Department of Entomology, Chaudhary Charan Singh Haryana Agricultural University, Hisar 125004, Haryana, India, p. 382.

Diarisso, N.Y., 1997. Spikelet Flowering Time and Morphology as Causes of Sorghum Resistance to Sorghum Midge (Diptera: Cecidomyiidae) (Ph.D. thesis). Texas A&M University, College Station.

Doerge, R.W., 2002. Mapping and analysis of quantitative trait loci in experimental populations. Nature Reviews Genetics 3, 43–52.

Dogget, H., Starks, K.J., Eberhart, S.A., 1970. Breeding for resistance to the sorghum shoot fly. Crop Science 10, 528–531.

Dogget, H., 1972. Breeding for resistance to sorghum shoot fly in Uganda. In: Jotwani, M.G., Young, W.R. (Eds.), Control of Sorghum Shoot Fly. Oxford and IBH Publishing Co., New Delhi, pp. 192–201.

Dogget, H., Majisu, B.N., 1965. Sorghum breeding research. Annual Report East African Agriculture for Research Organization 70–79.

Emani, C., Sunilkumar, G., Rathore, K.S., 2002. Transgene silencing and reactivation in sorghum. Plant Science 162, 181–192.

FaMing, W., SongMin, Z., YongHua, H., YuTao, S., ZhenYing, D., Yang, G., KunPu, Z., Xin, L., DaWei, L., Jin-Hua, C., DaoWen, W., 2013. Efficient and fine mapping of *RMES1* conferring resistance to sorghum aphid *Melanaphis sacchari*. Molecular Breeding 31, 777–784.

Faris, M.A., Lira, M.A., Veiga, A.F.S.L., 1979. Stability of sorghum midge resistance. Crop Science 19, 577–580.

Finch, S., Billiald, H., Collier, R.H., 2003. Companion planting-do aromatic plants disrupt host-plant finding by the cabbage root fly and the onion fly more effectively than non-aromatic plants? Entomologia Experimentalis et Applicata 109, 183–195.

Fisher, K.S., Wilson, F.L., 1975. Studies of grain production in *Sorghum bicolor* (L. Moench). III. The relative importance of assimilate supply, grain growth capacity and transport system. Australian Journal of Agricultural Research 26, 11–23.

Franzmann, B.A., Butler, D.G., 1993. Compensation for reduction in seed set due to sorghum midge (*Contarinia sorghicola*) in midge-susceptible and midge-resistant sorghum (*Sorghum bicolor*). Australian Journal of Experimental Agriculture 33, 193–196.

Franzmann, B.A., 1988. Components of resistance to sorghum midge in grain sorghum. In: Proceedings of Ninth Australian Plant Breeding Conference, pp. 277–278.

Gibson, P.T., Maiti, R.K., 1983. Trichomes in segregating generations of sorghum matings. I. Inheritance of presence and density. Crop Science 23, 73–75.

Gikonyo, N.K., Lwande, W., Hassanali, A., 2003. Periodicity in the quantity and ratio of pheromone components in volatile emissions from virgin females of the spotted stalk borer moth *Chilo partellus* (Swinhoe). East and Central African Journal of Pharmaceutical Sciences 6, 36–42.

Girijashankar, V., Sharma, H.C., Sharma, K.K., Swathisree, V., Sivarama Prasad, L., Bhat, B.V., Royer, M., San Secundo, B., Lakshmi Narasu, M., Altosaar, I., Seetharama, N., 2005. Development of transgenic sorghum for insect resistance against the spotted stem borer (*Chilo partellus*). Plant Cell Reports 24, 513–522.

Goud, J.V., Anahosur, K.H., Kulkarni, K.A., 1983. Breeding for multiple resistance in sorghum. In: Proceedings of National Seminar on Breeding Crop Plants for Resistance to Pests and Diseases, 25–27th May, 1983, Tamil Nadu Agricultural University, Coimbatore, p. 3.

Gray, S.J., Zhang, S., Rathus, C., Lemaux, P.G., Godwin, I.D., 2004. Development of sorghum transformation: organogenic regeneration and gene transfer methods. In: Seetharama, N., Godwin, I.D. (Eds.), Sorghum Tissue Culture and Transformation. Oxford Publishers, New Delhi, India, pp. 35–43.

Hagi, H.M., 1984. Gene Affects for Resistance to Stem Borer (*Chilo partellus* Swinhoe) in Sorghum (*Sorghum bicolour* (L.) Moench) (M.Sc. thesis). Andhra Pradesh Agricultural Universiy, Hyderabad, Andhra Pradesh, India.

Halalli, M.S., Gowda, B.T.S., Kulkarni, K.A., Goud, J.V., 1983. Evaluation of advanced generation progenies for resistance to shoot fly in sorghum. Indian Journal of Genetics and Plant Breeding 43, 291–293.

Hamilton, R.I., Subramanian, B., Reddy, M.N., Rao, C.H., 1982. Compensation in grain yield components in a panicle of rainfed sorghum. Annals of Applied Biology 101, 119–125.

Hansson, B.S., Blackwell, A., Hallberg, E., Löfqvist, J., 1995. Physiological and morphological characteristics of the sex pheromone detecting system in male corn stemborers, *Chilo partellus* (Lepidoptera: Pyralidae). Journal of Insect Physiology 41, 171–178.

Harris, K.M., 1962. Lepidopterous stemborers of cereals in Nigeria. Bulletin of Entomological Research 53, 139–172.

Harvey, T.L., Hackerott, H.L., 1974. Effects of greenbugs on resistant and susceptible sorghum seedlings in the field. Journal of Economic Entomology 67, 377–380.

Henzell, R.G., Gillieron, W., 1973. Effect of partial and complete panicle removal on the rate of death of some *Sorghum bicolor* genotypes under moisture stress. Queensland Journal of Agriculture and Animal Science 30, 291–299.

Henzell, R.G., Hare, B.W., 1996. Sorghum breeding in Australia – public and private. In: Proceedings of the Third Australian Sorghum Conference, Occasional Publication No. 93, pp. 159–171.

Henzell, R.G., Bregman, R.L., Page, F.D., 1980. Transference of sorghum midge resistance into agronomically acceptable lines. In: Proceedings of the Australian Agronomy Conference, Lawes, April 1980, p. 233.

Henzell, R.G., Franzmann, B.A., Brengman, R.L., 1994. Sorghum midge research in Australia. International Sorghum and Millets Newsletter 35, 41–47.

Ho, D.T., Reddy, K.V.S., 1983. Monitoring of lepidopterous stem-borer population by pheromone and light traps. Insect Science and Its Application 4, 19–23.

Hofte, H., Whiteley, H.R., 1989. Insecticidal crystal proteins of *Bacillus thuringiensis*. Microbiological Reviews 53, 242–255.

ICAR, 2002–07. Research Achievements of AICRPs on Crop Sciences. Indian Council of Agricultural Research, Directorate of Information & Publications of Agriculture, Krishi Anusandhan Bhavan, New Delhi, 129 pp.

ICRISAT, 1992. The Medium Term Plan, vol. 2. International Crops Research Institute for the Semi-Arid Tropics, Patancheru 502 324, AP, India.

IIMR, 2016. Indian Institute of Millets Research – Research achievements 2015–16. http://www.millets.res.in/ra15-16.php.

Jadhav, R., Jadhav, L.D., 1978. Studies on preliminary screening of some sorghum hybrids and varieties against earhead midge (*Contarinia sorghicola* Coq.). Journal of the Maharashtra Agricultural Universities 3, 187–188.

Jadhav, S.S., Mote, U.N., 1986. Interaction of shoot fly egg laying damage with resistant and susceptible sorghum varieties. Sorghum Newsletter 29, 69.

JinHua, C., XueYan, X., Li, Z., RongGai, L., GuoQing, L., YaoWu, L., 2006. Analysis of the resistance gene to the sorghum aphid, *Melanaphis sacchari*, with SSR marker in *Sorghum bicolor*. Acta Prataculturae Sinica 15, 113–118.

Johnson, J.W., Rosenow, D.T., Teetes, G.L., 1973. Resistance to the sorghum midge in converted exotic sorghum cultivars. Crop Science 13, 754–755.

Jordan, D.R., Tao, Y.Z., Godwin, I.D., Henzell, R.G., Copper, M., McIntyre, C.L., 1998. Loss of genetic diversity associated with selection for resistance to sorghum midge in Australian sorghum. Euphytica 102, 1–7.

Jotwani, M.G., Srivastava, K.P., 1970. Studies on sorghum lines resistant against shoot fly *Atherigona soccata* Rond. Indian Journal of Entomology 32, 1–3.

Jotwani, M.G., Young, W.R., 1972. Recent developments in chemical control of insect pests of sorghum. Pages 377–398. In: Rao, N.G.P., House, L.R. (Eds.), Sorghum in Seventies. Oxford and IBH Publishing Co., New Delhi, India.

Jotwani, M.G., Chandra, D., Young, W.R., Sukhani, T.R., Saxena, P.N., 1971. Estimation of avoidable losses caused by insect complex on sorghum hybrid CSH-1 and percentage increase in yield over treated control. Indian Journal of Entomology 33, 375–383.

Jotwani, M.G., 1976. Host plant resistance with special reference to sorghum. In: Proceedings of the National Academy of Science India vol. 46 (1–2), pp. 42–48.

Jotwani, M.G., 1978. Investigations on Insect Pests of Sorghum and Millets with Special Reference to Host Plant Resistance Final Technical Report, 1972–77. Indian Agricultural Research Institute, New Delhi, p. 114.

Jotwani, M.G., 1982. Factors reducing sorghum yields: insect pests. Pages 251–255. In: Sorghum in the Eighties. Proceedings of the International Symposium on Sorghum, 27 Nov 1981, LCRISAT Center, India, vol. 1. International Crops Research Institute for the Semi Arid Tropics, Patancheru A.P. 502 324, India.

Kalaisekar, A., Subbarayudu, B., Shyam Prasad, G., Bhagwat, V.R., 2010. Effect of cannibalism and interspecific competition on predatory potential of *Cheilomenes sexmaculata* (F.) and *Coccinella transversalis* F. in sorghum. In: National Seminar on Research and Development in Millets Held on 12th November, 2010 Organized by Directorate of Sorghum Research, Hyderabad, p. 97.

Kalaisekar, A., Patil, J.V., Shyam Prasad, G., Bhagwat, V.R., Padmaja, P.G., Subbarayudu, B., Srinivasa Babu, K., Rahman, Zeenat, 2013. Time-lapse tracing of biological events in an endophytic schizophoran fly, *Atherigona soccata* Rondani (Diptera: Muscidae). Current Science 105 (5), 695–701.

Kalaisekar, A., Patil, J.V., Shyam Prasad, G., Subbarayudu, B., Padmaja, P.G., Bhagwat, V.R., Srinivas Babu, K., 2014. Cereal shoot flies: basic aspects of ecofriendly management. In: Ramamurthy, V.V., Subramanian, S. (Eds.), Compilation of Invited Lectures Presented in National Symposium on Entomology as a Science and IPM as a Technology – the Way Forward, 14–15 November 2014 Held at the College of Horticulture and Forestry, Central Agricultural University, Pasighat – 791002, Arunachal Pradesh, India. Entomological Society of India, New Delhi, India.

Kamala, V., Sharma, H.C., Manahor Rao, D., Varaprasad, K.S., Bramel, P.J., 2009. Wild relatives of sorghum as sources of resistance to sorghum shoot fly, *Atherigona soccata*. Plant Breeding 128 (2), 137–142.

Kamatar, M.Y., Salimath, P.M., 2003. Morphological traits of sorghum associated with resistance to shoot fly, *Atherigona soccata* Rondani. Indian Journal of Plant Protection 31, 73–77.

Karanjkar, R.R., Chundurwar, R.D., Borikar, T., 1992. Correlations and path analysis of shoot fly resistance in sorghum. Journal of Maharashtra Agricultural Universities 17, 389–391.

Kausalya, K.G., Nwanze, K.F., Reddy, Y.V.R., Nwilene, F.E., 1997. .A simple head cage technique for monitoring sorghum midge (Diptera: Cecidomyiidae). International Journal of Pest Management 43, 35–38.

Kayumbo, H.Y., 1976. Incidence of the spotted borer *Chilo* sp. on sorghum in Morogoro region, Tanzania. In: Presented at the Sixth East African Cereals Research Conference, 13–18 May 1976, Morogoro, Tanzania.

Kfir, R., 1994. Attempts at biological control of the stem borer *Chilo partellus* (Swinhoe) (Lepidoptera: Pyralidae) in South Africa. African Entomology 2 (1), 67–68.

Khan, M.Q., Rao, A.S., 1956. The influence of the black ant (*Camponotus compressus* F.) on the incidence of two homopterous crop pests. Indian Journal of Entomology Society 18, 199–200.

Khurana, A.D., Verma, A.N., 1982. Amino acid contents in sorghum plants, resistance/susceptible to stemborer and shootfly. Indian Journal of Entomology 44, 184–188.

Khurana, A.D., Verma, A.N., 1983. Some biochemical plant characters in relation to susceptibility of sorghum to stemborer and shootfly. Indian Journal of Entomology 45, 29–37.

Kishore, P., 2001. Resistance to shoot fly, Atherigona soccata Rondani and stem borer, Chilo partellus (Swinhoe) in new germplasm of sorghum. Journal of Entomological Research 25 (4), 273–282.

Kishore, G.K., Pande, S., Rao, J.N., 2002. Field evaluation of plant extracts for control of late leaf spot in groundnut. International Arachis Newsletter 22, 46–48.

Kishore, P., 1987. Key pest on sorghum, pearl millet and smaller millets and their management. pp. 243–259. In: Mathur, Y.K., Bhatnagar, A.K., Pandey, N.D., Srivastava, J.P. (Eds.), Recent Advances in Entomology Gopala Prakashan, Parade, Kanpur (India).

Kishore, P., 1989. Chemical control of stem borers, ICRISAT (International Crop Research Institute for Semi Arid Tropics) 1989. In: International Workshop on Sorghum Stem Borers, 17–20 Nov 1987, ICRISAT Center, India. ICRISAT, Patancheru, AP. 502 324, India, pp. 73–80.

Kishore, P., 1992. Evaluation of twin resistance sources amongst advance generation derivatives in sorghum to shoot fly, *Atherigona soccata* Rondani and stem borer, *Chilo partellus* (Swinhoe). Journal of the Entomological Research Society 16 (3), 236–241.

Kishore, P., 1994. Development of new dual purpose sorghum germplasm showing resistance to shoot fly, *Atherigona soccata* Rondani and stem borer, *Chilo partellus* (Swinhoe). Journal of the Entomological Research Society 18 (3), 279–281.

Kogan, M., Ortman, E.E., 1978. Antixenosis – a new term proposed to replace Painter's 'non-preference' modality of resistance. Bulletin of the Entomological Society of America 24, 175–176.

Kumar, D.A., Channaveerswami, A.S., 2015. Pre and post emergence control measures for shootfly incidence and its influence on seed yield of little millet (*Panicum sumatrense*). Journal of Experimental Zoology 18 (2), 811–814.

Kumar, L.V., Prabhuraj, A., 2005. *Monomorium fassulatus* Emery: a predatory ant on sorghum shoot bug *Perigrinus maidis* (Ashmead). Insect Environment 11, 40–41.

Kumar, P.A., Sharma, R.P., Malik, V.S., 1996. The insecticidal proteins of *Bacillus thuringiensis*. Advances in Applied Microbiology 42, 1–43.

Kumar, L.V., Prabhuraj, A., Navyashree, K., 2005. An ecto-parasitic mite on sorghum shoot bug, *Perigrinus maidis* (Ashmead) (Homoptera: Delphacidae). Insect Environment 11, 74–75.

Kumar, P.A., 2003. Insect pest-resistant transgenic crops. In: Upadhyay, R.K. (Ed.), Advances in Microbial Control of Insect Pest. Kluwer Academic, New York, pp. 71–142.

Lal, G., Sukhani, T.R., 1982. Antibiotic effects of some resistant lines of sorghum on post-larval development of *Chilo partellus* Swinhoe. The Indian Journal of Agricultural Sciences 52, 127–129.

Lux, S.A., Hassanali, A., Lwande, W., Njogu, F.N., 1994. Proximity of release points of pheromone components as a factor confusing males of the spotted stem borer, *Chilo partellus*, approaching the trap. Journal of Chemical Ecology 20, 2065–2075.

Maiti, R.K., Bidinger, F.R., Seshu Reddy, K.V., Gibson, P., Davies, J.C., 1980. Nature and Occurrence of Trichomes in Sorghum Lines with Resistance to the Sorghum Shoot Fly. Joint Progress Report of Sorghum Physiology/Sorghum Entomology, vol. 3. ICRISAT, Patancheru.

Maiti, R.K., Prasada Rao, K.E., Raju, P.S., House, L.R., 1984. The glossy trait in sorghum: its characteristics and significance in crop improvement. Fields Crops Research 9, 279–289.

Managoli, S.P., 1973. An attack of shoot bug/pundalouya-bug (*Peregrinus maidis*) on rabi jowar in dry tract of Bijapur district. Farmers Journal (India) 16–17.

Mate, S.N., Phadanwis, B.A., Mehetre, S.S., 1988. Studies on growth and physiological factors in relation to shoot fly attack on sorghum. Indian Journal of Agricultural Research 22, 81–84.

Meksongsee, B., Chawanapong, M., 1985. Sorghum insect pests in South East Asia. In: Proceedings of the International Sorghum Entomology Workshop, July 15–24, 1984, Texas A&M University, College Station, Texas, USA. International Crops Research Institute for the Semi-Arid Tropics (ICRISAT), Patancheru, Andhra Pradesh 502 324, India, pp. 57–64.

Miles, C.M., Wayne, M., 2008. Quantitative trait locus (QTL) analysis. Nature Education 1 (1), 108.

Moreno, C.R., Racelis, A.E., 2015. Attraction, Repellence, and Predation: Role of Companion Plants in Regulating Myzus persicae (Sulzer) (Hemiptera: Aphidae) in Organic Kale Systems of South Texas. Southwestern Entomologist 40 (1), 1.

Mote, U.N., Shahane, A.K., 1993. Studies on varietal reaction of sorghum to delphacid, aphid, and leaf sugary exudation. Indian Journal of Entomology 55, 360–367.

Mote, U.N., Shahane, A.K., 1994. Biophysical and biochemical characters of sorghum varieties contributing resistance to delphacid, aphid and leaf sugary exudations. Indian Journal of Entomology 56, 113–122.

Mote, U.N., Bapat, D.R., Kadam, J.R., Ghule, B.D., 1985. Studies on the causes of 'leaf sugary malady' on sorghum and its control. In: Paper Presented in the Annual AICSIP Workshop, May 6–8, 1985, Punjabrao Krishi Vidyapeeth, Akola, Maharashtra, India 5 pp.

Mote, U.N., 1984. Sorghum species resistant to shootfly. Indian Journal of Entomology 46 (2), 241–243.

Murthy, T.K., Harinarayana, G., 1989. Insect-pests of small millets and their management in India. In: Seetharam, A., Riley, K.W., Harinarayana, G. (Eds.), Small Millets in Global Agriculture. Proc. First Int. Small Millets Workshop, Bangalore. Oxford and IBH Publishing Co. Pvt. Ltd., New Delhi, pp. 255–270.

Napompeth, B., 1973. Ecology and Population Dynamics of the Corn Planthopper, *Peregrinus maidis* (Ashmead) (Homoptera: Delphacidae) in Hawaii (Ph.D. thesis). University of Hawaii, Honolulu, Hawaii. 257 pp.

Narayana, N.D., 1975. Characters contributing to sorghum shoot fly resistance. Sorghum Newsletter 18, 21.

Narwal, R.P., 1973. Silica bodies and resistance to infection in jowar (*Sorghum vulgare* Pers.). Agra University Journal of Research (Science) 22, 17–20.

Natarajan, K., Chelliah, S., 1985. Studies on the sorghum grain midge, *Contarinia sorghicola* Coquillet, in relation to environmental influence. Tropical Pest Management 31, 276–285.

Natarajan, K., Chelliah, S., 1986. Studies of mechanism of resistance in sorghum accessions to the grain midge, Contarinia sorghicola coquillett. Insect Science and Its Application 7, 751–755.

Nesbitt, B.F., Beevor, P.S., Hall, D.R., Lester, R., Davies, J.C., Reddy, K.V.S., 1979. Components of the sex pheromone of the female spotted stalk borer, Chilo partellus (Swinhoe) (Lepidoptera:Pyralidae): identification and preliminary field trials. Journal of Chemical Ecology 5 (1), 153–163.

Ni, X., Wilson, J.P., Buntin, G., 2009. Differential responses of forage pearl millet genotypes to chinch bug (Heteroptera: Blissidae) feeding. Journal of Economic Entomology 102, 1960–1969.

Nimbalkar, V.S., Bapat, D.R., 1992. Inheritance of shoot fly resistance in sorghum. Journal of Maharashtra Agricultural Univerisities 17, 93–96.

Nwanze, K.F., Reddy, Y.V.R., Soman, P., 1990. The role of leaf surface wetness in larval behaviour of the sorghum shoot fly, *Atherigona soccata*. Entomologia Experimentalis et Applicata 56, 187–195.

Nwanze, K.F., Pring, R.J., Sree, P.S., Butler, D.R., Reddy, Y.V.R., Soman, P., 1992. Resistance in sorghum to the shoot fly, *Atherigona soccata*: epicuticular wax and wetness of the central whorl leaf of young seedlings. Annals of Applied Biology 120, 373–382.

Ogwaro, K., Kokwaro, E.D., 1981. Development and morphology of the immature stages of the sorghum shoot fly, *Atherigona soccata* Rondani. Insect Science and Its Application 1, 365–372.

Omori, T., Agrawal, B.L., House, L.R., 1983. Componential analysis of the factors influencing shoot fly resistance in sorghum (*Sorghum bicolor* L. (Moench)) (*Atherigona soccata*). Japan Agricultural Research Quarterly 17, 215–218.

Padmaja, P.G., Woodcock, C.M., Bruce, T.J.A., 2010. Electrophysiological and behavioral responses of sorghum shoot fly, *Atherigona soccata*, to sorghum volatiles. Journal of Chemical Ecology 36 (12), 1346–1353.

Page, F.D., 1979. Resistance to sorghum midge (*Contarinia sorghicola* Coquillett) in grain sorghum. Australian Journal of Experimental Agriculture and Animal Husbandry 19, 97–101.

Painter, R.H., 1951. Insect Resistance in Crop Plants. University Press of Kansas, Lawrence.

Pant, N.C., Gupta, P., Nayar, J.K., 1960. Physiological studies of *Chilo zonellus* Swinh. A pest on maize crop. I. Growth on artificial diets. Proceedings of the National Institute of Sciences, India 26B, 379–383.

Parmar, G.M., Juneja, R.P., Mungra, K.D., 2015. Management of shoot fly and stem borer on pearl millet crop. International Journal of Plant Protection 8 (1), 104–107.

Parolin, P., Bresch, C., Desneux, N., Brun, R., Bout, A., Boll, R., et al., 2012. Secondary plants used in biological control: a review. International Journal of Pest Management 58, 91–100.

Patil, S.B., Jaiarao, K., Khot, R.S., 1996. Efficacy of *Trichogramma chilonis* (Ishii) in the management of early shoot borer of sugarcane, *Chilo infuscatellus* (Snell). Bharatiya Sugar 22 (4), 43–44 3 ref.

Patil, S.S., Narkhede, B.N., Barbate, K.K., 2006. Effects of biochemical constituents with shoot fly resistance in sorghum. Agricultural Science Digest 26, 79–82.

Polaszek, A., Walker, A.K., 1991. The *Cotesia flavipes* species complex: parasitoids of cereal stemborers in the tropics. Redia 74, 335–341.

Ponnaiya, B.W.X., 1951. Studies in the genus *Sorghum*. I. Field observations on sorghum resistance to the insect pest, *Atherigona indica* M. Madras Agricultural Journal 21, 96–117.

Pradhan, S., Prasad, S.K., 1955. Correlation between the degree of damage due to *Chilo zonellus* Swin. and the yield of jowar grain. Indian Journal of Entomology 15, 136–137.

Pradhan, S., et al., 1971. Investigations on Insect Pests of Sorghum and Millets final technical report (1965–70). Indian Agricultural Research Institute, New Delhi, India.

Raina, A.K., Thindwa, H.K., Othieno, S.M., Cork-Hill, R.T., 1981. Resistance in sorghum to the sorghum shoot fly: larval development and adult longevity and fecundity on selected cultivars. Insect Science and Its Application 2, 99–103.

Raina, A.K., 1981. Deterrence of repeated oviposition in sorghum shootfly *Atherigona soccata*. Journal of Chemical Ecology 7, 785–790.

Raina, A.K., 1985. Mechanisms of resistance to shootfly in sorghum. A review pp. 131–136. In: Proceedings of International Sorghum Entomology Workshop, 15–21 July, 1984 Texas A&M University, College Station, TX USA. ICRISAT, Patancheru, A.P, 502 324, India.

Rakshit, S., Gomashe, S.S., Ganapathy, K.N., Elangovan, M., Ratnavathi, C.V., Seetharama, N., Patil, J.V., 2012. Morphological and molecular diversity reveal wide variability among sorghum Maldandi landraces from India. Journal of Plant Biochemistry and Biotechnology 21 (2), 145–156.

Rana, B.S., Jotwani, M.G., Rao, N.G.P., 1981. Inheritance of host plant resistance to the sorghum shootfly. Insect Science and Its Application 2 (1–2), 105–109.

Rana, B.S., Singh, B.U., Rao, V.J.M., Reddy, B.B., Rao, N.G.P., 1984. Inheritance of stemborer resistance in sorghum. Indian Journal of Genetics and Plant Breeding 44, 7–14.

Rana, B.S., Singh, B.U., Rao, N.G.P., 1985. Breeding for shoot fly and stemborer resistance in sorghum. In: Proceedings of the International Sorghum Entomology Workshop, 15–21 July 1984, Texas A&M University, College Station, TX, USA, pp. 347–360.

Ranjekar, P.K., Patankar, A., Gupta, V., Bhatnagar, R., Bentur, J., Kumar, P.A., 2003. Genetic engineering of crop plants for insect resistance. Current Science 84, 321–329.

Rao, S.B.P., Rao, D.V.N., 1956. Studies on the sorghum shoot borer fly, *Atherigona indica* Malloch (Anthomyiidae-Diptera) at Siruguppa. Mysore Agricultural Journal 31, 158–174.

Rao, N.G.P., Rana, B.S., Jotwani, M.G., 1977. Host plant resistance to major insect pests of sorghum. In use of induced mutations for resistance of crop plants to insects. In: Proceedings, FAO/IAEA Advisory Group, Dakar (Senegal), I.A.E.A, Vienna.

Ratnadass, A., Ryckewaert, P., Claude, Z., Nikiema, A., Pasternak, D., Woltering, L., Thunes, K., Zakari-Moussa, O., Hale, C., 2011. New ecological options for the management of horticultural crop pests in Sudano-Sahelian agroecosystems of West Africa. Acta Horticulturae 917, 85–91.

Ratnadass, A., Chantereau, J., Coulibaly, M.F., Cilas, C., 2002. Inheritance of resistance to the panicle-feeding (Eurystylus oldi) and the sorghum midge (stenodiplosis sorghicola) in sorghum. Euphytica 123, 131–138.

Reddy, K.V.S., Skinner, J.D., Davies, J.C., 1981. Attractants for *Atherigona* spp. including the sorghum shootfly, *Atherigona soccata* Rond. (Muscidaer Diptera). Insect Science and Its Applications 2, 83–86.

Rossetto, C.J., Goncalves, W., Diniz, J.L.M., 1975. Resistencia de variedade AF-28 a mosca do sorgo, *Contarinia sorghicola*, no ausencia de outras variedades. Anais Da Sociedade Entomologica Do Brazil 4, 16–20.

Rossetto, C.J., Nagai, V., Overman, J., 1984. Mechanisms of resistance in sorghum variety AF-28 to *Contarinia sorghicola* (Diptera: Cecidomyiidae). Journal Economic Entomology 77, 1439–1440.

Rossetto, C.J., 1977. Tipos de resistencia de sorgo *Sorghum bicolor* (L.) Moench, a *Contarinia sorghicola* (Coquillet, 1898) (In Pt.). Thesis. Universidade Estudual Paulista, Jaboticabal, Sao Paulo, Brazil. 34 pp.

Sajjanar, G.M., 2002. Genetic Analysis and Molecular Mapping of Components of Resistance to Shoot Fly (*Atherigona soccata* Rond.) in Sorghum (*Sorghum bicolor* (L.) Moench) (Ph.D. thesis). Department of Genetics and Plant Breeding, University of Agricultural Sciences, Dharwad, Karnataka, India, p. 265.

Salin, K.P., Kanaujia, K.R., 1999. Facultative deposition of oviposition deterring and stimulating pheromones by shoot fly, *Atherigona soccata* Rondani. Journal of Insect Science 12, 63–64.

Sandhu, G.S., Young, W.R., 1974. Chemical control of sorghum shoot fly in India. Pesticide 8, 105–135.

Sanjana, R.P., Patil, J.V., 2015. Genetic Enhancement of Rabi Sorghum: Adapting the Indian Durras. Academic Press, p. 248.

Satish, K., Srinivas, G., Madhusudhana, R., Padmaja, P.G., Nagaraja Reddy, R., Murali Mohan, S., Seetharama, N., 2009. Identification of quantitative trait loci for resistance to shoot fly in sorghum [Sorghum bicolor (L.) Moench]. Theoretical and Applied Genetics 119, 1425–1439.

Satish, K., Madhusudhana, R., Padmaja, P.G., Seetharama, N., Patil, J.V., 2012. Development, genetic mapping of candidate gene-based markers and their significant association with the shoot fly resistance quantitative trait loci in sorghum [*Sorghum bicolor* (L.) Moench]. Molecular Breeding 30, 1573–1591.

Saxena, K.N., 1990. Mechanisms of resistance/susceptibility of certain sorghum cultivars to the stem borer *Chilo partellus*: role of behaviour and development. Entomologia Experimentalis et Applicata 55, 91–99.

Saxena, K.N., 1992. Larval development of *Chilo partellus* (Swinhoe) (Lepidoptera: Pyralidae) on artificial diet incorporating leaf tissues of sorghum lines in relation to their resistance or susceptibility. Applied Entomology and Zoology 27, 325–330.

Schnepf, E., Crickmore, N., Van Rie, J., Lereclus, D., Baum, J., Feitelson, J., Zeigler, D.R., Dean, D.H., 1998. *Bacillus thuringiensis* and its pesticidal crystal proteins. Microbiology and Molecular Biology Reviews 62, 775–806.

Sharma, H.C., Davies, J.C., 1988. Insect and other animal pest of millets. International Crops Research Institute for the Semi-Arid Tropics, Patancheru, Andhra Pradesh 502 324, India, p. 86.

Sharma, H.C., Franzmann, B.A., 2001. Orientation of sorghum midge, *Stenodiplosis sorghicola*, females (Diptera: Cecidomyiidae) to color and host-odor stimuli. Journal of Agricultural and Urban Entomology 18 (4), 237–248.

Sharma, H.C., Lopez, V.F., 1990. Mechanisms of resistance in sorghum to head bug, *Calocoris angustatus*. Entomologia Experimentalis et Applicata 57 (3), 285–294.

Sharma, H.C., Vidyasagar, P., 1992. Orientation of males of sorghum midge, *Contarinia sorghicola* to sex pheromones from virgin females in the field. Entomologia Experimentalis Applicata 64, 23–29.

Sharma, H.C., Vidyasagar, P., 1994. Antixenosis component of resistance to sorghum midge, *Contarinia sorghicola* Coq. in *Sorghum bicolor* (L.) Moench. Annals of Applied Biology 124, 495–507.

Sharma, H.C., Nwanze, K.F., 1997. Mechanisms of resistance to insects in sorghum and their usefulness in crop improvement. Information Bulletin No. 45. International Crops Research Institute for the Semi-Arid Tropics (ICRISAT), Patancheru, Andhra Pradesh, India, p. 56.

Sharma, G.C., Jotwani, M.G., Rana, B.S., Rao, N.G.P., 1977. Resistance to the sorghum shoot fly, *Atherigona soccata* (Rondani) and its genetic analysis. Journal of Entomological Research 1, 1–12.

Sharma, H.C., Leuschner, K., Vidyasagar, P., 1990a. Factors influencing oviposition behaviour of the sorghum midge, *Contarinia sorghicola* Coq. Annals of Applied Biology 116, 431–439.

Sharma, H.C., Leuschner, K., Vidyasagar, P., 1990b. Componental analysis of the factors influencing resistance to sorghum midge, *Contarinia sorghicola* Coq. Insect Science and Its Application 11 (6), 889–898.

Sharma, H.C., Taneja, S.L., Leuschner, K., Nwanze, K.F., 1992. Techniques to screen sorghums for resistance to insect pests. In: Information Bulletin No. 32. International Crops Research Institute for the Semi-Arid Tropics, Patancheru, p. 48.

Sharma, H.C., Vidyasagar, P., Subramanian, V., 1993. Antibiosis component of resistance in sorghum to sorghum midge, *Contarinia sorghicola*. Annals of Applied Biology 123, 469–483.

Sharma, H.C., Nwanze, K.F., Subramanian, V., 1997. Mechanisms of resistance to insects and their usefulness in sorghum improvement. In: Sharma, H.C., Singh, F., Nwanze, K.F. (Eds.), Plant Resistance to Insects in Sorghum. Patancheru, International Crop Research Institute for Semi Arid Tropics, Patancheru, pp. 81–100.

Sharma, H.C., Sankaram, A.V.B., Nwanze, K.F., 1999a. Utilization of natural pesticides derived from neem and custard apple in integrated pest management. In: Singh, R.P., Saxena, R.C. (Eds.), *Azadirachta indica* A. Juss., pp. 199–213.

Sharma, H.C., Ananda Kumar, P., Seetharama, N., Hari Prasad, K.V., Singh, B.U., 1999b. Role of transgenic plants in pest management in sorghum. In: Symposium on Tissue Culture and Genetic Transformation of Sorghum, 23–28 Feb 1999. ICRISAT Center, Patancheru, Andhra Pradesh, India.

Sharma, H.C., Kumari, A.P.P., Reddy, D.R.R., 2000a. Components of resistance to the sorghum head bug, *Calocoris angustatus*. Crop Protection 19 (6), 385–392.

Sharma, H.C., Mukuru, S.Z., Gugi, H., King, S.B., 2000b. Inheritance of resistance to sorghum midge and leaf disease in Sorghum in Kenya. International Sorghum and Millets Newsletter 41, 37–42 ISSN:1023-487X.

Sharma, H.C., Sharma, K.K., Seetharama, N., Ortiz, R., 2000c. Prospects for using transgenic resistance to insects in crop improvement. Electronic Journal of Biotechnology 3 (2), 1–26 ISSN:0717-3458.

Sharma, H.C., Babu, B.S., Surender, A., Rao, R.D.V.J.P., Chakrabarty, S., Singh, S.D., Girish, G.A., 2000d. Sorghum germplasm from Thailand showing resistance to sugarcane aphid, *Melanaphis sacchari* Zehntner. Indian Journal of Plant Genetic Resources 13 (2), 186–187.

Sharma, H.C., Franzmann, B.A., Henzell, R.G., 2002. Mechanisms and diversity of resistance to sorghum midge, *Stenodiplosis sorghicola*. Euphytica 124, 1–12.

Sharma, H.C., Taneja, S.L., Kameswara Rao, N., Prasada Rao, K.E., 2003. Evaluation of sorghum germplasm for resistance to insect pests. In: Information Bulletin No. 63. International Crop Research Institute for the semi-Arid Tropics, Patancheru 502324, Andhra Pradesh, India. 177 pp.

Sharma, H.C., Bhagwat, V.R., Padmaja, P.G., 2008. Techniques to screen sorghums for resistance to insect pests. In: Reddy, B.V.S., Ramesh, S., Ashok Kumar, A., Gowda, C.L.L. (Eds.), Sorghum Improvement in the New Millennium. ICRISAT, India, pp. 31–49.

Sharma, H.C., 1985. Screening for sorghum midge resistance and resistance mechanisms. In: Proceeding of the International Sorghum Entomology Workshop, Texas A&M University and ICRISAT, pp. 89–95.

Sharma, H.C., 1993. Host plant resistance to insects in sorghum and its role in integrated pest management. Crop Protection 12, 11–34.

Sharma, M.L., 1996. Efficiency of virgin female trap for monitoring and trapping of male moth of *Chilo partellus* on sorghum. Indian Journal of Entomology 58, 349–353.

Sharma, H.C., 1997. Plant resistance to insects: basic principles. In: Sharma, H.C., Singh, F., Nwanze, K.F. (Eds.), Plant Resistance to Insects in Sorghum. International Crop research Institute for Semi Arid Tropics, Patancheru, pp. 24–31.

Shekharappa, Kulkarni, K.A., 2003. Cultural practices for the management of stem borer, *Chilo partellus* (Swinhoe) in sorghum. Indian Journal of Plant Protection 31 (1), 134–136.

Shivankar, V.I., Ram, S., Gupta, M.P., 1989. Tolerance in some sorghum germplasm to shoot fly (*Atherigona soccata* Rondani). Indian Journal of Entomology 51, 593–596.

Shyamsunder, J., Parameswarappa, R., Nagaraja, H.K., Kajjari, N.B., 1975. A new genotype in sorghum resistant to midge (*Contarinia sorghicola* Coq.). Sorghum Newsletter 20, 68.

Siddiqui, K.H., Sarup, P., Panwar, V.P.S., Marwaha, K.K., 1977. Evolution of basegradients to formulate artificial diets for mass rearing of *Chilo partellus* (Swinhoe). Journal of the Entomological Research Society 1, 117–131.

Singh, S.P., Batra, G.R., 2001. Effect of neem formulations on shoot fly, *Atherigona soccata* (Rondani) oviposition and infestation in forage sorghum. Haryana Agricultural University Journal of Research 31, 9–11.

Singh, S.P., Jotwani, M.G., 1980a. Mechanisam of reisistance in sorghum shoot fly. III. Biochemical basis of resistance. Indian Journal of Entomology 42, 551–566.

Singh, S.P., Jotwani, M.G., 1980b. Mechanisms of resistance in sorghum to shoot fly. II. Antibiosis. Indian Journal of Entomology 42, 5–60.

Singh, J.P., Marwaha, K.K., 1996. Persistance and residual toxicity of insecticides on maize leaves/whorls against freshly hatched larvae of maize stalk bore, *Chilo parrellus* (Swinhoe.). Indian Journal of Entomology 57, 213–218.

Singh, R., Narayana, K.L., 1978. Influence of different varieties of sorghum on the biology of sorghum shoot fly. Indian Journal of Agricultural Sciences 48, 8–12.

Singh, B.U., Rana, B.S., 1984. Influence of varietal resistance on oviposition and larval development of stalk-borer *Chilo partellus* Swinhoe and its relationship to field resistance in sorghum. Insect Science and Its Application 5, 287–296.

Singh, B.U., Rana, B.S., 1986. Resistance in sorghum to the shoot fly, *Atherigona soccata* Rondani. Insect Science and Its Application 7, 577–587.

Singh, B.U., Rana, B.S., 1992. Stability of resistance to corn planthopper, *Peregrinus maidis* (Ashmead) in sorghum germplasm. Insect Science and Its Application 13, 251–263.

Singh, V.S., Shankar, K., 2000. Screening of sorghum genotypes for their reaction to stemborer and shoot fly. Indian Journal of Entomology 62 (1), 34–36.

Singh, B.U., Sharma, H.C., 2002. Natural enemies of sorghum shoot fly. *Atherigona soccata* Rondani (Diptera: Muscidae). Biocontrol Science and Technology 12 (3), 307–323.

Singh, S.P., Verma, A.N., 1988. Antibiosis mechanism of resistance to stem borer, *Chilo partellus* (Swinhoe) in sorghum. Insect Science and Its Application 9, 579–582.

Singh, S.R., Vedamoorthy, G., Thoi, V.V., Jotwani, M.G., Young, W.R., Balan, J.S., Srivastava, K.P., Snadhu, G.S., Krishnanada, N., 1968. Resistance to Stemborer, Chilo Zonellus (Swinhoe) and Stem Fly, *Atherigona socaata* Rond. In the World Sorghum Collection in India. Memoirs of the Entomological Society of India, p. 79.

Singh, S.P., Jotwani, M.G., Rana, B.S., Rao, N.G.P., 1978. Stability of host plant resistance to sorghum shoot fly, *Atherigona soccata* (Rond.). Indian Journal of Entomology 40, 376–383.

Singh, P., Unnithan, G.C., Delobel, A.G.L., 1983. An artificial diet for sorghum shootfly larvae. Entomologia Experimentalis et Applicata 33, 122–124.

Singh, O., Shrivastava, O.S., Verma, S.N.P., Rathore, J.S., 1990. Field evaluation of some kodo selections for resistance to shoot fly (*Atherigona* sp.). Indian Journal of Entomology 52, 191–194.

Singh, B.U., Padmaja, P.G., Seetharama, N., 2004. Stability of biochemical constituents and their relationships with resistance to shoot fly, *Atherigona soccata* (Rondani) in seedling sorghum. Euphytica 136, 279–289.

Soto, P.E., Laxminarayan, K., 1971. A method for rearing sorghum shoot fly. Journal of Economic Entomology 64, 553.

Soto, P.E., 1972. Mass rearing of sorghum shoot fly and screening for host plant resistance under greenhouse conditions. In: Jotwani, M.G., Young, W.R. (Eds.), Control of Sorghum Shoot Fly. Oxford & IBH, New Delhi, India, pp. 137–138.

Soto, P.E., 1974. Ovipositional preference and antibiosis in relation to resistance to sorghum shoot fly. Journal of Economic Entomology 67, 165–167.

Sukhani, T.R., Jotwani, M.G., 1979. A simple artificial diet for mass-rearing of the sorghum shootfly, *Atherigona soccata* (Rondani). Bulletin of Entomology 20, 61–66.

Sukhani, T.R., Jotwani, M.G., 1980. Efficacy of mixtures of carbofuran treated and untreated sorghum seed for the control of shoot fly, *Atherigona soccata* (Rondani). Journal of Entomological Research 4, 186–189.

Sukhani, T.R., 1986. Insect pest management in sorghum. Plant Protection Bulletin 38, 57–62.

Swaine, G., Wyatt, C.A., 1954. Observations on the sorghum shoot fly. East African Agricultural and Forestry Journal 20, 45–48.

Swarup, V., Chaugale, D.S., 1962. A preliminary study of resistance of stem borer, *Chilo zonellus* (Swinhoe) infestation in sorghum (*Sorghum vulgare* Pers.). Current Science 31, 163–164.

Taley, Y.M., Thakare, K.R., 1979. Biology of seven new hymenopterous parasitoids of *Atherigona soccata* (Rondani). Indian Journal of Agricultural Sciences 49, 344–354.

Tan, W.Q., Li, S.M., Guo, H.P., Gao, R.P., 1985. A study of the inheritance of aphid resistance in sorghum. Shanxi Nongye Kexue (Shanxi Agricultural Science) 8, 12–14.

Taneja, S.L., Henry, V.K., 1993. Chemical control of sorghum shoot-fly: dosage, method and frequency of insecticide application in India. Crop Protection 12 (1), 74–78.

Taneja, S.L., Leuschner, K., 1985. Resistance screening and mechanisms of resistance in sorghum to shoot fly. In: Proceedings of the International Sorghum Entomology Workshop, 15–21 July 1984, Texas A&M University, College Station, Texas, USA, pp. 115–129.

Tao, Y.Z., Hardy, A., Drenth, J., Henzell, R.G., Franzmann, B.A., Jordan, D.R., 2003. Identifications of two different mechanisms for sorghum midge resistance through QTL mapping. Theoretical and Applied Genetics 107, 116–122.

Thimmaiah, G., Panchabhavi, K.S., Desai, K.S.M., Usman, S., Kajjari, N.B., 1973. Chemical control of sorghum shoot fly (*Atherigona varia soccata* Rondani) (Diptera: Anthomyiidae) in Mysore state. Indian Journal of Agricultural Sciences 43, 294–298.

Torto, B., Hassanali, A., Saxena, K.N., 1990. Chemical aspects of *Chilo partellus* feeding on certain sorghum cultivars. Insect Science and Its Application 11, 649–655.

Torto, B., Addae-Mensah, I., Moreka, L., 1992. Antifeedant activity of *Piper guineense* Schum and Thonn. amides against larvae of the sorghum stem borer *Chilo partellus* (Swinhoe). Insect Science and Its Application 13, 705–708.

Unnithan, G.C., Reddy, K.V.S., 1985. Oviposition and infestation of sorghum shoot fly, *Atherigona soccata* Rondani on certain sorghum cultivars in relation to their relative resistance and susceptibility. Insect Science and Its Application 6, 409–412.

Unnithan, G.C., Saxena, K.N., 1990. Population monitoring of *Chilo partellus* (Swinhoe) (Lepidoptera: Pyralidae) using pheromone traps. Insect Science and Its Application 11, 795–805.

Unnithan, G.C., Saxena, K.N., 1991. Pheromonal trapping of *Chilo partellus* (Swinhoe) (Lepidoptera: Pyralidae) moths in relation to male population density and competition with females. Applied Entomology and Zoology 26, 17–28.

Usman, S., 1973. Efficacy of granular insecticide for sorghum shoot fly control. In: Jotwani, M.G., Young, W.R. (Eds.), Control of Sorghum Shoot Fly. Oxford and IH Company, New Delhi, pp. 252–261.

van den Berg, J., van der Westhuizen, M.C., 1997. *Chilo partellus* (Lepidoptera: Pyralidae) moth and larval response to levels of antixenosis and antibiosis in sorghum inbred lines under laboratory conditions. Bulletin of Entomoogical Research 87, 541–545.

van den Berg, J., Bronkhorst, L., Mgonja, M., Obilana, A.B., 2005. Resistance of sorghum varieties to the shoot fly, *Atherigona soccata* Rondani (Diptera: Muscidae) in Southern Africa. International Journal of Pest Management 51, 1–5.

Van Frankenhuyzen, K., 2009. Insecticidal activity of *Bacillus thuringiensis* crystal proteins. Journal of Invertebrate Pathology 101, 1–16.

van Rensburg, N.J., van Hamburg, H., 1975. Grain sorghum pests: an integrated control approach. In: Proceedings of the First Congress of Entomological Society of Southern Africa, 1975, pp. 151–162.

van Rensburg, N.J., 1973. Notes on the occurrence and biology of the sorghum aphid in South Africa. Journal of the Entomological Society of South Africa 36, 293–298.

Van den Berg, J., Nur, A.F., 1998. Chemical control. In: African Cereal Stem Borers: Economic Importance, Taxonomy, Natural Enemies and Control. CABI, Wallingford, UK, pp. 319–332.

Vedamoorthy, G., Thobbi, V.V., Matai, B.H., Young, W.R., 1965. Indian Journal of Agricultural Sciences 35, 14–28.

Venkateswaran, 2003. Diversity Analysis and Identification of Sources of Resistance to Downy Mildew, Shoot Fly and Stemborer in Wild Sorghums (Ph.D. thesis). Department of Genetics, Osmania University, Hyderabad, Andhra Pradesh, India.

Verma, O.P., Bhanot, J.P., Verma, A.N., 1992. Development of *Chilo partellus* (Swinhoe) on pest resistant and susceptible sorghum cultivars. Journal of Insect Science 5, 181–182.

Vijay Kumar, L., Prabhuraj, A., 2007. Bio-efficacy of new seed treatment chemicals against shoot fly and shoot bug on rabi sorghum in North Karnataka. Indian Journal of Crop Science 2 (1).

Vinayan, M.T., Hash, C.T., Deshpande, S.P., Sharma, H.C., Mohanasundaram, K., Robin, S., 2011. Importance of sorghum chromosome sbi-07 in controlling resistance to spotted stem borer [*Chilo partellus* (Swinhoe)]. http://www.ncpgr.cn/papers/abs_usa_19th.html (accessed on 11th July 2015).

Visarada, K.B.R.S., Padmaja, P.G., Saikishore, N., Pashupatinath, E., Kanti Meena, R.,S.V., Seetharama, N., 2007. Genetic transformation of sorghum for resistance to stemborer. In: Poster Presented at Agriculture Science Congress, Coimbatore, Feb 2007.

Waquil, J.M., Teetes, G.L., Peterson, G.C., 1986. Adult sorghum midge (Diptera: Cecidomyiidae) nonpreference for resistant hybrid sorghum. Journal of Economic Entomology 79, 455–458.

Wiseman, B.R., Duncan, R.R., Widstrom, N.W., 1988. Registration of SGIRL-MR-3 and SGIRL-MR-4 midge resistant sorghum germplasms. Crop Science 28, 202–203.

Woodhead, S., Taneja, S.L., 1987. The importance of the behaviour of young larvae in sorghum resistance to Chilo partellus. Insect Science and Its Applications 45, 47–54.

Wuensche, A.L., 1980. An Assessment of Plant Resistance to the Sorghum Midge, *Contarinia sorghicola*, in Selected Lines of *Sorghum bicolour* (Ph.D. thesis). Texas A&M University, College Station, TX, USA. 193 pp.

Zein el Abdin, A.M., 1981. Review of sorghum shoot fly research in the Sudan. Insect Science and Its Application 2, 55–58.

Zhang, M., Tang, Q., Chen, Z., Liu, J., Cui, H., Shu, Q., Xia, Y., Altosaar, I., 2009. Genetic transformation of Bt gene into sorghum (*Sorghum bicolor* L.) mediated by *Agrobacterium tumefaciens*. Chinese Journal of Biotechnology 25 (3), 418–423.

Zhao, Z.Y., Cai, T., Tagliani, L., Miller, M., Wang, N., Pang, H., Rudert, M., Schroeder, S., Hondred, D., Seltzer, J., Pierce, D., 2000. Agrobacterium-mediated sorghum transformation. Plant Molecular Biology 44, 789–798.

Zhong, H., Wang, W., Sticklen, M., 1998. In vitro morphogenesis of *Sorghum bicolor* (L.) Moench: efficient plant regeneration from shoot apices. Journal of Plant Physiology 153, 719–726.

Zhu, L., Lang, Z.-Hong, Li, G.-Ying, H.E, K.-Lai, YUE, T.-Qing, Zhang, J., Huang, Da-fang, 2011. Introduction of Bt cry1Ah gene into sweet sorghum (*Sorghum bicolor* L. Moench) by *Agrobacterium tumefaciens*-mediated transformation. Scientia Agricultura Sinica 44 (10).

Zimmerman, E.C., 1948. Insects of Hawaii, vol. 4. University Hawaii Press, Honolulu Hawaii, USA. 263 pp.

Zongo, J.O., Vincent, C., Stewart, R.K., 1993. Effects of neem seed kernel extracts on egg and larval survival of the sorghum shoot fly, *Atherigona soccata* Rondani (Dipt., Muscidae). Journal of Applied Entomology 115, 363–369.

FURTHER READING

Chikkarugi, N.M., Balikai, R.A., 2011. Response of sorghum genotypes in shoot pest nursery to major pests. Research Journal of Agriculture Sciences 2 (1), 21–25.

Farid, A., Khan, M.I.N., Khan, A., Khattak, S.U.K., Alamzeb, Sattar, A., 2007. Study on maize stem borer, *Chilo partellus* (Swin.) in Peshawar valley. Pakistan Journal of Zoology 9 (2), 127–131.

Rana, B.S., Murthy, B.R., 1971. Genetic analysis of resistance to stem borer in sorghum. Indian Journal of Genetics & Plant Breeding 31 (3), 521–529.

Sharma, G.C., Rana, B.S., 1983. Resistance to the sorghum shoot fly, *Atherigona soccata* (Rond.) and selection for antibiosis. Journal of Entomological Research 7 (2), 133–138.

Sharma, G.C., Rana, B.S., 1985. Genetics of ovipositional non-preference and dead-heart formation governing shootfly resistance in sorghum. Journal of Entomological Research 9 (1), 104–105.

Sharma, H.C., Ratnadass, A., 2000. Colour variation in the African sorghum head bug. International Sorghum and Millets Newsletter 41, 42–43.

Singh, A., Jain, A., Sarma, B.K., et al., 2014. Rhizosphere competent microbial consortium mediates rapid changes in phenolic profiles in chickpea during *Sclerotium rolfsii* infection. Microbiological Research 169 (5–6), 353–360.

Singh, B.U., Rana, B.S., Reddy, B.B., Rao, N.G.P., 1983. Host plant resistance to stalk-borer, *Chilo partellus* Swin., in sorghum. Insect Science and Its Application 4 (4), 407–413.

Songa, J.M., Overholt, W.A., Mueke, J.M., Okello, R.O., 2002. Regional distribution of lepidopteran stemborers and their parasitoids among wild grasses in the semi-arid Eastern Kenya. African Crop Science Journal 10 (2), 183–194.

Index

Printed in the United States
By Bookmasters